AN INTRODUCTION TO PRACTICAL BIOCHEMISTRY

AN INTRODUCTION TO PRACTICAL BIOCHEMISTRY

Arun Rastogi

ANMOL PUBLICATIONS PVT. LTD.
NEW DELHI-110 002 (INDIA)

ANMOL PUBLICATIONS PVT. LTD.
Regd. Office: 4360/4, Ansari Road, Daryaganj,
New Delhi-110002 (India)
Tel.: 23278000, 23261597, 23286875, 23255577
Fax: 91-11-23280289
Email: anmolpub@gmail.com
Visit us at: www.anmolpublications.com

Branch Office: No. 1015, Ist Main Road, BSK IIIrd Stage
IIIrd Phase, IIIrd Block, Bangalore-560 085 (India)
Tel.: 080-41723429 • Fax: 080-26723604
Email: anmolpublicationsbangalore@gmail.com

An Introduction to Practical Biochemistry

First Edition, 2010

ISBN 978-81-261-4500-3

PRINTED IN INDIA

Printed at Mehra Offset Press, Delhi.

Contents

Preface — *vii*

1. Protein as Coenzymes — 1
2. Peptides — 23
3. Carbohydrates as Biomolecule — 98
4. Vitamins Biosynthesis — 169
5. Chemistry of Sugars — 214
6. Biosynthesis of Amino Acids — 263
7. Chemistry of Antibiotics — 277

Index — 290

Preface

Biochemistry is the chemistry of life. Biochemists study the elements, compounds and chemical reactions that are controlled by enzymes and take place in all living organisms. The findings of biochemistry are used in many areas, from genetics to molecular biology and from agriculture to medicine.

Biochemistry processes concern not only chemists and biochemists, but also physicists and physicians. Over the past 20 years there have been considerable changes in the way we consider biochemistry. In view of the results obtained in chemistry and biochemistry, biochemistry appears to be much more than a simple tuning of the properties of the molecule itself, and important remote effects have been highlighted. It is now obvious that the inversion about either a double bond or a restrained single bond generates extensive changes.

It is an important component of all disciplines of biology. In the era of multidisciplinary approach, the basic techniques in biochemistry and molecular biology are much needed by the students of botany, zoology, microbiology, biotechnology, fisheries, veterinary, pharmacology, physiology, medicine, genetics, agriculture and allied subjects.

Author

Chapter 1

Protein as Coenzymes

Proteins are present in every cell of humans, animals, plant tissues, tissue fluids and in micro organisms. They account for about 50% of the dry weight of a cell. The term protein is derived from the Greek word *proteios* meaning holding first place or rank in living matter.

MEDICAL AND BIOLOGICAL IMPORTANCE

Proteins perform wide range of essential functions in mammalians.

- Proteins are involved in the transport of substances in the body.
  Example: Haemoglobin transports oxygen.
- Enzymes which catalyze chemical reactions in the body are proteins.
- Proteins are involved in defence function. They act against bacterial or viral infection.
  Example: Immunoglobulins.
- Hormones are proteins. They control many biochemical events.
  Example: Insulin.
- Some proteins have role in contraction of muscles.
  Example: Muscle proteins.
- Proteins are involved in the gene expression. They control gene expression and translation.
  Example: Histones.
- Proteins serve as nutrients. Proteins are also involved in storage function.
  Examples: Casein of milk, Ferritin that stores iron.

- Proteins act as buffers.
  Example: Plasma proteins.
- Proteins function as anti-vitamins.
  Example: Avidin of egg.
- Proteins are infective agents.
  Example: Prions which ca9. use mad cow disease are proteins.
- Some toxins are proteins.
  Example: Enterotoxin of cholera microorganism.
- Some proteins provide structural strength and elasticity to the organs and vasc'ılar system.
  Example: Collagen and elastin of bone matrix and ligaments.
- Some proteins are components of structures of tissues.
  Example: α-keratin is present in hair and epidermis.

In order to understand how these substances though they are all proteins play such diverse functions their structures, and composition must be explored.

CHEMICAL NATURE OF PROTEINS

All proteins are polymers of aminoacids. The aminoacids in proteins are united through "Peptide" linkage. Sometimes proteins are also called as polypeptides because they contain many peptide bonds.

PROPERTIES OF PROTEINS

- Proteins have high molecular weight, *e.g.*, the lactalbumin of milk molecular weight is 17000 and pyruvate dehydrogenase molecular weight is 7×10^6.
- Proteins are colloidal in nature.
- Proteins have large particle size.
- Different kinds of proteins are soluble in different solvents.
- Proteins differ in their shape.
- Some proteins yield amino acids only on hydrolysis where as others produce amino acids plus other types of molecules.

- *Charge properties*: Charge of a protein depends on the surroundings like amino acids. So, by changing the pH of surroundings the charge of protein can be altered. This property is used for separation of proteins.

  Isoelectric point: Proteins have characteristic isoelectric points. At the isoelectric point its net charge is zero because the number of positive charges are equal to number of negative charges.

  So proteins are insoluble or have minimum solubility at isoelectric point. This property is used for the isolation of casein from milk. The isoelectric point for casein is 4.6. If the pH of the surrounding is raised above the isoelectric point, the protein is negatively charged *i.e.*, it exists as anion. Likewise, if the pH of the surrounding is lowered, the protein is positively charged *i.e.*, it exist as cation. Further, proteins do not move in an electrical field at isoelectric point like amino acids. However, if the pH of the medium is raised or lowered protein moves towards anode or cathode respectively. This property is exploited for the separation of proteins.
- *Proteins act as buffers*: Since proteins are amphoteric substances, they act as buffers. Hemoglobin (Hb) of erythrocytes and plasma proteins are important buffers. Hb accounts for 60% of buffering action with in erythrocytes and plasma proteins contributes to 20% of buffering action of blood.

CLASSIFICATION OF PROTEINS

There is no single universally satisfactory system of protein classification so far.

- One system classifies proteins according to their composition or structure.
- One system classifies them according to solubility.
- One system classifies them according to their shape.
- Classification of proteins based on their function also found in literature.

Classification of Proteins Based on their Composition

Proteins are divided into three major classes according to their structure.

- *Simple proteins*: Simple proteins are made up of amino acids only. On hydrolysis, they yield only amino acids.
  Examples: Human plasma albumin, Trypsin, Chymotrypsin, pepsin, insulin, soyabean trypsin inhibitor and ribonuclease.
- *Conjugated proteins*: They are proteins containing non-protein part attached to the protein part. The non-protein part is linked to protein through covalent bond, non-covalent bond and hydrophobic interaction. The non-protein part is loosely called as prosthetic group. On hydrolysis, these proteins yield non-protein compounds and amino acids.
  Conjugated protein - Protein + Prosthetic group
  The conjugated proteins are further classified into subclasses based on prosthetic groups.

Table. Different classes of conjugated proteins

	Subclass	Prosthetic group	Examples	Type of linkage
1.	Lipoproteins	Lipids	Various classes of Lipoproteins. Lipovitellin of egg	Hydrophobhic Interaction
2.	Glycoproteins	Carbohydrates	Immunoglobulins of blood, Egg albumin	Covalent
3.	Phosphoproteins	Phosphorus	Casein of milk, vitellin of egg yolk	Covalent
4.	Nucleoproteins	Nucleic acids	Chromatin, Ribosomes	Non-covalent
5.	Hemoproteins/ Chromoproteins	Heme	Hemoglobin, Myoglobin, Cytochromes	Non-covalent
6.	Flavoproteins	Flavin nucleotides FMN, FAD	Succinate Dehydrogenase	Covalent
7.	Metalloproteins	Iron	Ferritin, Cytochromes	Non-covalent
8.	Visual pigments	Retinal	Rhodopsin	Covalent

- *Derived proteins*: As the name implies this class of proteins are formed from simple and conjugated proteins. There are two classes of derived proteins.
  - *Primary derived proteins:* They are formed from natural proteins by the action of heat or alcohol etc. The peptide bonds are not hydrolysed. They are synonymous with denatured proteins.
    Example: Coagulated proteins like cooked-egg

albumin.

– *Secondary derived proteins:* They are formed from partial hydrolysis of proteins.
Examples: Proteoses, peptone, gelatin, and peptides.

Proteins Classification According to their Solubility

- *Albumins*: Soluble in water and salt solutions.
Examples: Albumin of plasma, egg albumin and lactalbumin of milk.
- *Globulins:* Sparingly soluble in water but soluble in salt solutions.
Examples: Globulins of plasma, ovoglobulins of egg, lactoglobulin of milk.
- *Glutelins:* Soluble in dilute acids and alkalies.
Examples: Glutenin of wheat, oryzenin of rice, zein of maize.
- *Protamins:* Soluble in ammonia and water.
Examples: Salmine from salmon fish, sturine of sturgeon.
- *Histones:* Soluble in water and dilute acids.
Example: Histones present in chromatin.
- *Prolamines:* Soluble in dilute alcohol and insoluble in water and alcohol.
Examples: Gliadin of wheat, zein of corn.
- *Clero proteins*: Insoluble in water and dilute acids and alkalies.
Examples: Collagen, elastin and keratin.

Classification of Proteins Based on Shape

Proteins are divided into two classes based on their shape.

- *Globular proteins:* Polypeptide chain(s) of these proteins are folded into compact globular (Spherical) shape.
Examples: Haemoglobin, myoglobin, albumin, lysozyme, chymotrypsin.
- *Fibrous proteins:* Poly peptide chains are extended along one axis.
Examples: α-keratin, β-keratin, collagen and elastin.

PROTEIN STRUCTURE

Since proteins are built from amino acids by linking them in linear fashion, it may be viewed as proteins having long chain like structures. However, such arrangement is unstable and polypeptide or protein folds to specific shape known as *conformation,* which is more stable.

Various stages involved in the formation of final conformation from linear chain are divided into four levels or orders of protein structure.

Primary Structure

The linear sequence of amino acid residues in a polypeptide chain is called as primary structure. Generally disulfide bonds if any are also included in the primary structure. *Bonds responsible for the maintenance of primary structure* are mainly peptide bonds and *disulfide* bonds. Both of them are covalent bonds.

Primary Structure of Insulin

This protein consist of two polypeptide chains A and B. The two chains are covalently linked by disulfide bonds. The A chain has N-terminal glycine and C-terminal aspargine. The B chain has phenylalanine and alanine as N-and C-terminal residues, respectively. Insulin is a hormone and its molecular weight is 5,700.

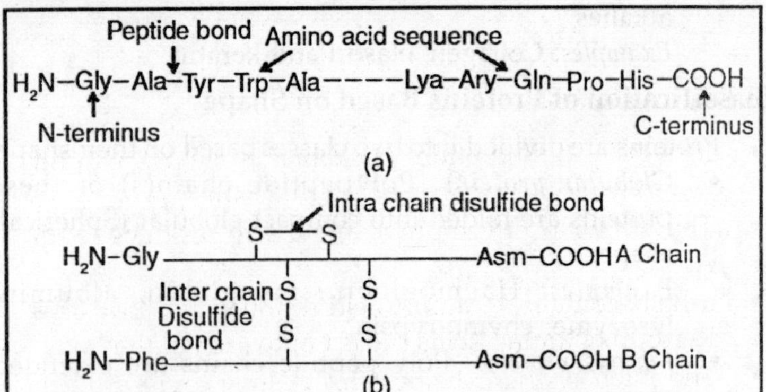

Fig. (a) Structure of a protein (b)Insulin primary structure

Secondary Structure

Folding of polypeptide chain along its long axis is called as secondary structure of protein. Folding of polypeptide chain can be ordered, disordered or random. Secondary structure is often referred as conformation. So, proteins has ordered secondary structure or conformation and random or disordered secondary structure or conformation.

Ordered Conformation of Polypeptides

The polypeptide chain of some proteins may exist in highly ordered conformation. The conformation is maintained by *hydrogen bonds* formed between peptide residues.

Hydrogen Bond

It is a weak ionic interaction between positively charged hydrogen atom and negatively charged atoms like oxygen, nitrogen, sulfur etc. It is indicated with broken lines (—-).

There are two types of ordered secondary structure observed in proteins.

- The polypeptide chain of keratin, which is present in hair, nails, epidermis of the skin is arranged as α-*Helix*: α-letter is given to this type of structure because it was first ordered structure noticed in proteins.
- Polypeptide chain of keratin, which is present in silk fibroin and spider web is arranged in β-*pleated* sheet. The β-letter is given because it was observed later.

Main Features of α-Helix

- In a-helix polypeptide, backbone is tightly wound round (coiled) long axis of the molecule.
- The distance between two amino acid residues is 1.5 Å.
- α-helix contain 3.6 amino acid residues per turn. The R-group of amino acids project outwards of the helix.
- The pitch of the α-helix is 5.4 Å long and width is 5.0 Å.

- The α-helix is stabilized by intra chain hydrogen bonds formed between –N–H groups and –C=O groups that are four residues back, *i.e.,* –N–H group of a 6th peptide bond is hydrogen bonded to –C=O group of 2nd peptide bond.

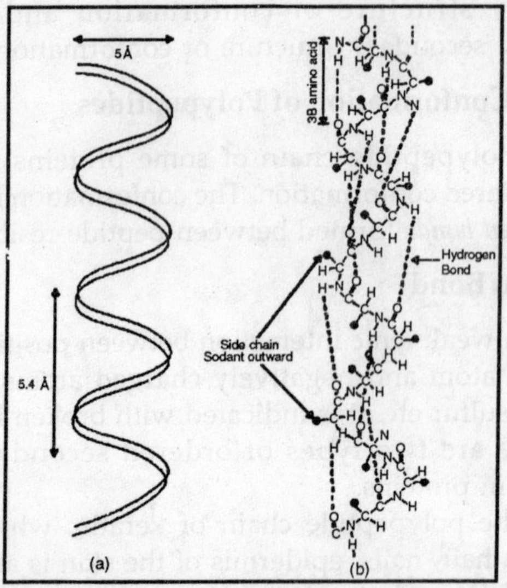

Fig. (a) Right handed a-helix

- Each peptide bond participates in the hydrogen bonding. This gives maximum stability to α-helix.
- α-helix present in most fibrous proteins is right handed. The right handed α-helix is more stable than the left handed helix.
- α-helix is hydrophobic in nature because of intra chain hydrogen bonds.
- An α-helix forms spontaneously since it is the most stable conformation of polypeptide chain.
- Some amino acids act as terminators for α-helix. *Example:* Proline.
- Aromatic amino acids stabilizes α-helix.
- Charged and hydrophobic amino acids destabilize α-helix.
- Content of α-helix varies from protein to protein.

β-Pleated Sheet Features

- In β-pleated sheet, the polypeptide chain is fully extended.4
- In β-pleated sheet, polypeptide chains line up side by side to form sheet. The side chains are above or below the plane of the sheet.
- From 2 to 5, adjacent strands of polypeptides may combine and form these structure.
- When the adjacent polypeptide chains run in same direction (N to C terminus) the structure is termed as parallel β-pleated sheet.
- When the adjacent polypeptide chains run in opposite direction the structure is termed as anti-parallel β-pleated sheet.
- The β-pleated sheet is stabilized by inter chain hydrogen bonds.
- β-keratin contains anti parallel β-pleated sheet.
- Both parallel and anti-parallel β-pleated sheet occur in ot function of proteins as those of helices and β-pleated sheet.It accumulates in the CNS.

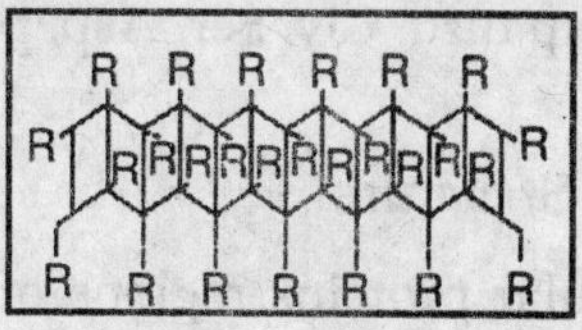

Fig. β Pleated Sheet Showing Arrangement of R groups all the R-groups Project Above (solid line) or Below (broken- line)the Plane.

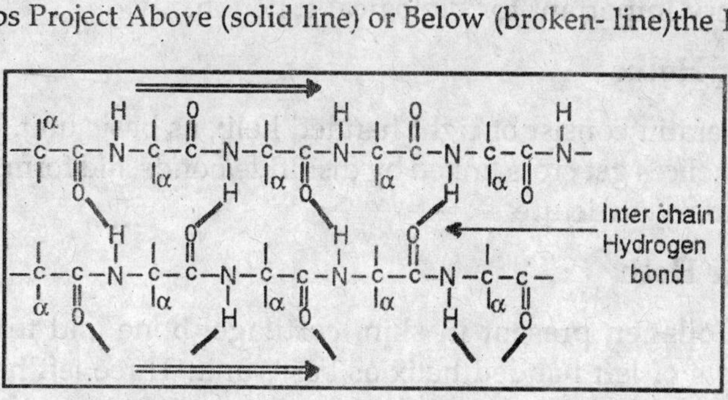

Fig. Parallel β-structure Showing Interchain Hydrogen Bonds

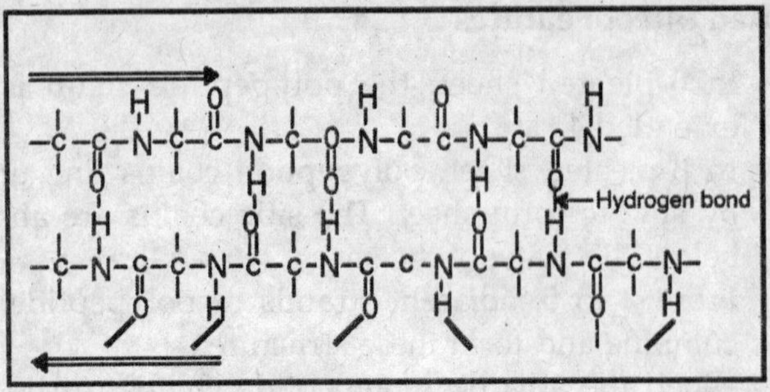

Fig. Anti-Parallel β-Structure

Random Coil (Disordered) Conformation

Regions of proteins that are not organized as helices and pleated sheet are said to be present in andom coil conformation. These are also equally important for biological function of proteins as those of helices and pleated sheet turn or bends (Reverse Turn).

Hair pin turn of a polypeptide chain is called as turn. The change in the direction of a polypeptide chain is achieved by turn. turn connects anti parallel sheets. Usually four aminoacids make up turn. Gly, Ser, Asp, proline are involved in turns.

Super Secondary Structure

In some globular proteins regions of helix and pleated sheet join to form super secondary structure or motifs. They are very important for biological function.

Super Helix

keratin consist of right handed helix as basic unit. Three such helices get cross linked by disulfide bonds and form super secondary structure.

Triple Helix

Collagen present in skin, cartilage, bone and tendons consists of left handed helix as basic unit. Three left handed helices are wrapped around each other to right handed super secondary structure triple helix.

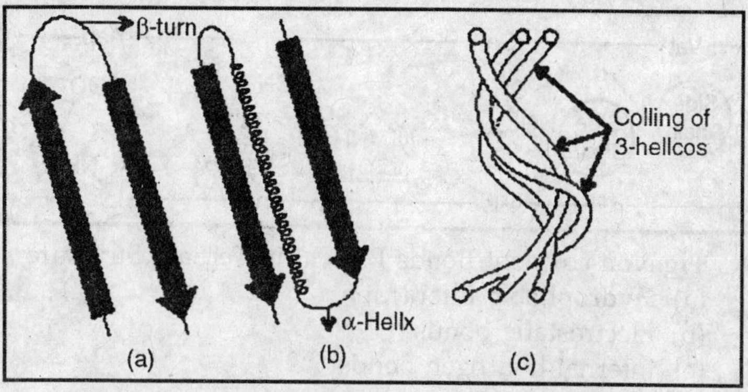

Fig. (a) Two Anti-parallel Chains are Joined by β-turn (b) Motif (c) Super Secondry Structure

Tertiary Structure

Three-dimensional folding of polypeptide chain is called as tertiary structure. It consists of regions of helices, pleated sheet, turns, motifs and random coil conformations. Interrelationships between these structures are also a part of tertiary structure. Tertiary structure of a protein is mainly stabilized by non-covalent bonds.

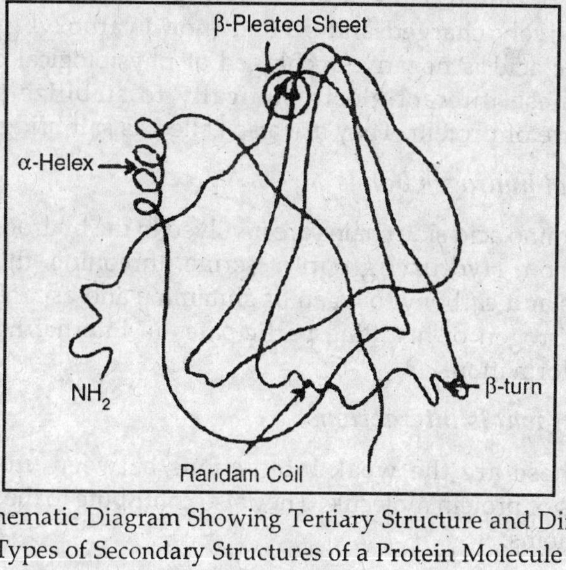

Fig. Schematic Diagram Showing Tertiary Structure and Different Types of Secondary Structures of a Protein Molecule

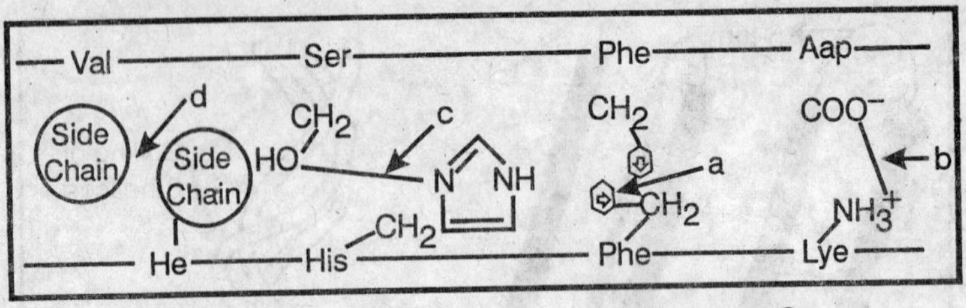

Fig.Non-covalent Bonds Present in Tertiary Structure

(a) Hydrophobic interaction
(b) Electrostatic bonds
(c) Internal hydrogen bonds
(d) Vander waal's interactions

Hydrophobic interactions

The non-polar side chains of neutral amino acids tend to associate in proteins. These are called as hydrophobic interactions. They play significant role in maintaining tertiary structure.

Electrostatic bonds

These bonds are formed between oppositely charged groups of amino acid side chains. The amino groups of lysine is positively charged and second (non-) carboxyl group of aspartic acid is negatively charged at physiological or body pH. These interact electrostatically to stabilize tertiary structure of protein. They are also called as salt bridges.

Internal hydrogen bonds

Amino acid side chains are involved in the hydrogen bond formation. Hydroxyl group of serine, threonine, the amino groups and carbonyl oxygen of glutamine and aspargine, the ring nitrogen of histidine participates in internal hydrogen bond formation.

Vander waals interactions

These are the weak interactions between uncharged groups of protein molecule. They also contribute to the stability of proteins.

Quaternary Structure

Proteins containing two or more polypeptide chains possess quaternary structure. These proteins are called as *oligomers*. The individual polypeptide chains are called as protomer, *monomers* or *subunits*. The protomers are united by forces other than covalent bonds. Occasionally, they may be joined by disulfide bonds.

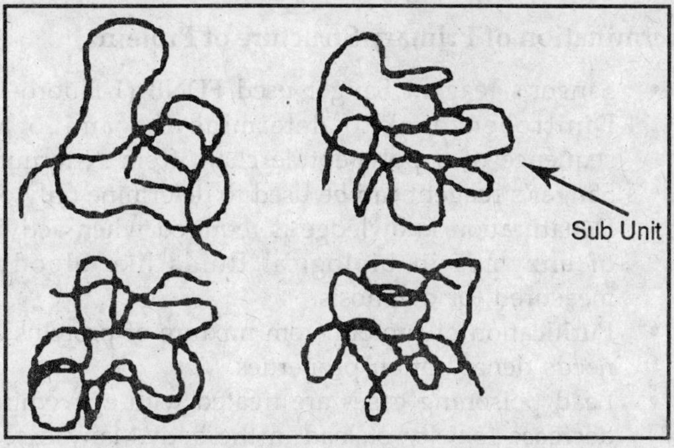

Fig. Quaternary Structure of a Tetramer

The most common oligomeric proteins contain 2 or 4 protomers and are termed dimmers and tetramers.

Forces that Stabilize these Aggregates (Assembles of Monomers) are:

- Hydrogen bonding
- Electrostatic interactions
- Hydrophobic interactions
- Vander waals interactions
- Disulfide bridges (in some proteins)

Examples:

- Haemoglobin consist of 4 polypeptide chains.
- Hexokinase contains 2 subunits.
- Pryuvate dehydrogenase contains 72 subunits.

Determination of Protein Structure

The primary structure of protein directs specific folding

(secondary structure) and its tertiary structure. If there is a change in one of the amino acids of protein, then conformation of polypeptide chain alters, which results change in biological function. Further, the sequence of amino acids in proteins that gives them their striking specific biological actions. Therefore knowledge of primary structure of a protein is required for the understanding of relationship of a protein's structure to its function at molecular level.

Determination of Primary Structure of Protein

- Sanger's reagent Sanger used FDNB (1-Fluoro-2, 4-Dinitrobenzene) to determine the amino acid sequence of a polypeptide chain from N-terminus. Sanger's reagent can be used to determine only one
- Denaturation knowledge is required when activities of enzymes in biological fluids like blood are measured for diagnosis.
- Purification of protein from mixture of proteins also needs denaturation properties.
- Lead poisoning cases are treated with egg white to decrease toxicity of lead in the body.Many cases of the process of denaturation is irreversible.

  Examples of Denaturation:
- When egg white is exposed to high temperature coagulum is formed because heat denatures egg albumin. The solubility of denatured protein is decreased.
- Formation of coagulum when albumin is exposed to high temperature.
- Heat treatment of trypsin results in loss of biological activity.
- Monellin is a dimeric protein has sweet taste. On denaturation the sweet taste is lost. Renaturation Though denaturation is irreversible in majority of the cases, in few cases, renaturation is observed.

  Example: Ribonuclease denatures on exposure to heat but come back to its native conformation when temperature is lowered.

PLASMA PROTEINS

Plasma is non-cellular portion of blood. The total plasma protein level ranges from 6-7 gm/dl. Plasma contains many structurally and functionally different proteins. Plasma proteins are divided into two categories.

- *Albumin:* Not precipitated by half-saturated ammonium sulfate.
- *Globulin:* Precipitated by half-saturated ammonium sulfate. The albumin constitutes over half of the total protein. Albumin level ranges from 3.5-5.5 gm/dl.

Globulin ranges from 2-3 gm/dl. After the age of 40, albumin gradually declines with an increase in globulins. Albumin is found to be simple protein and a single entity. But globulin has been found to contain many components.

Subglobulins are detected as bands on electrophoresis. They are globulins. Electrophoretic pattern is shown. The different plasma protein bands are semi-quantitated using densitometer.

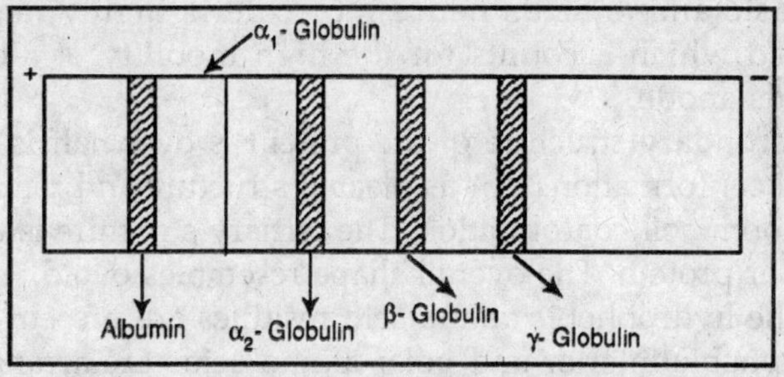

Fig. Eiectrophoretic Pattern of Plasma Proteins

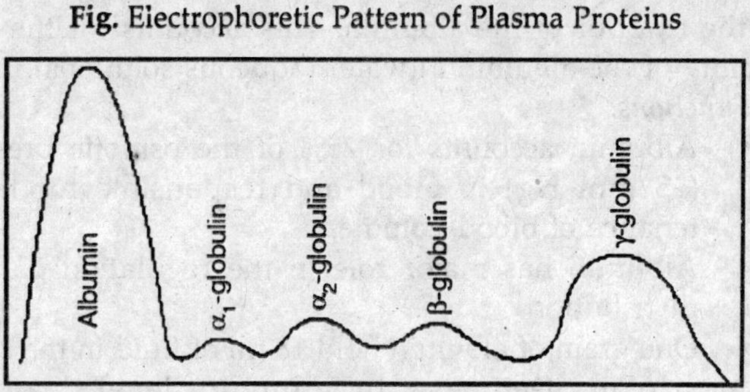

Fig. Densitometer Scan of Different Plasma Protein Bands

Characteristics of Plasma Proteins

- They are all glycoproteins except albumin. Sialic acid is the most important of all the sugars present in plasma proteins. Removal of sialic acid decreases the life span of plasma proteins.
- Each plasma protein has defined life span. The half life of albumin is 20 days and haptoglobin life span is 15 days.
- Liver is the sole source of albumin, prothrombin and fibrinogen. Most of the globulins are also of hepatic origin. globulins are derived from lymphocytes.

Albumin

Liver produces about 12 gms of albumin per day.

It consists of single polypeptide chain of 584 amino acid residues with a molecular weight of 66,300. Charged amino acids (glutamate, aspartate and lysine) make up a quarter of the total amino acid residues. The acidic residues out number the basic amino acids hence molecule is highly negative charged which accounts for the high mobility of albumin towards anode.

Secondary structure of the protein is over half is in the helical conformation. 15% as pleated structure and remaining in random coil conformation. The tertiary structure is that of globular protein. The overall shape resembles ovoid.

The hydrophobic amino acid residues are present in the hydrophobic interior and polar amino acids are arranged to face the exterior of the albumin. This accounts for the high solubility of the albumin in water (aqueous solutions).

Functions:

- Albumin accounts for 75% of the osmotic pressure (25 mm Hg) in blood and responsible for maintenance of blood volume.
- Albumin has major role in the regulation of fluid distribution.
- One gram of albumin hold 18 ml of fluid in the blood stream. Decrease in albumin level leads to accumulation of fluid which results in edema.

- It transports fatty acids from adipose tissue to liver. Albumin also binds many hydrophobic substances like bilirubin and several drugs. The binding of bilirubin is critical in neo-natal period.
- Albumin act as a reservoir for Ca^{2+} in plasma. About 40% of plasma calcium is bound to albumin.
- Albumin is also involved in the transport of thyroid hormones, glucocorticoids and sex steroids.
- Albumin function as protein source for peripheral tissues. Each day liver replaces about 12 gm of albumin taken up by peripheral tissues. In certain conditions like stress and starvation the turn over rate of albumin is increased. Albumin is in dynamic equilibrium.
- Albumin acts as a buffer.

1-Globulin: Mainly 1-antitrypsin. It is a protease inhibitor. It is the major component of 1-fraction and accounts more than 90%. It inhibits trypsin, chymotrypsin, elastase and neutral protease. The major function of 1-antitrypsin is the protection of pulmonary tissue and other tissues from the destructive action of proteases.

$$\underset{(\alpha_1\text{–AT}) \text{ or AAT}}{\alpha_1 - \text{antitrypsin}} \xrightarrow{\text{Elastase}} \underset{\text{(In active)}}{\alpha_1 - AT - \text{Elastase complex}} \rightarrow$$

$$\text{No proteolysis} \rightarrow \underset{\text{damage}}{\text{No lung}}$$

$$\text{Lung Tissue} \xrightarrow{\text{Elastase}} \text{Proteolysis} \rightarrow \text{Lung } \textit{damage}$$

Deficiency of α_1–antitrypsin results in emphysema

1-Acid glycoprotein (AAG): It is another major component of 1-globulins. It increases in plasma in inflammatory conditions.

Other components of 1-globulins are Lipoprotein: Functions in the transport of lipids (HDL). It transports cholesterol from extra hepatic tissue to liver.

Prothrombin: Blood clotting factor.

Retinolbinding protein: Transport of Vit A.

Thyroxine binding globulin: Transport of thyroxine.

1-Fetoprotein: It is present only in fetal serum. Its presence in non-foetal serum indicates primary carcinoma of liver. It is referred as tumour marker.

2-Globulins: The 2-fraction of globulins includes.

Haptoglobulin: It combines with haemoglobin in order to remove it from the circulation.

Kidney can not filter haemoglobin-haptoglobin complex because of its larger size.

2-Macroglobulin: It functions as protease inhibitor. It combines with proteases and facilitates their removal from circulation. It also binds with cytokines and involved in zinc transport.

Ceruloplasmin: A copper binding plasma protein and function as ferrooxidase and converts

$$Fe^{2+} \rightarrow Fe^{3+}$$

Erythropoietin: It is involved in erythropoiesis.

Pseu docholinesterase: It is only functional enzyme present in plasma. It hydrolyzes acetylcholine.

Transferrin: It accounts for about 60% of globulins. It is an iron transport protein.

Lipoproteins: Involved in the transport of cholesterol from live to extrahepatic tissue (LDL).

Complement-3: It is one of the member of complement system present in plasma. It is involved in phagocytosis.

Other globulins present in plasma are:

Fibrinogen: It is similar to globulins because it is precipitated by half saturation with ammonium sulfate. It is a fibrous or filamentous protein. It is the precursor of fibrin, the blood clotting substances.

Prealbumin: It is a component of globulin fraction. Though it is a globulin by nature it is named as prealbumin because it migrates ahead of a albumin in electrophoresis. It is a carrier of thyroxine, Vitamin A and binds calcium. Other blood clotting factors, plasminogen and several non-functional enzymes are also present in plasma.

Acute Phase Proteins or Reactants (APR)

- The concentration of these proteins increases markedly during acute inflammation.

- They are 1-antitrypsin, haptoglobin, ceruloplasmin, complement-3, fibrinogen and c-reactive protein. Their concentration increases in conditions like surgery, myocardial infraction, infections and tumours.
- Acute phase reaction is general to any infection. They all play part in complex defensive process of inflammation.
- The synthesis of these proteins by liver is triggered by interleukin at the site of injury.
- The plasma levels of these APR raises at different rates. The levels of c-reactive protein raises first followed by 1-antitrypsin. The level of complement-3 raises at the end.

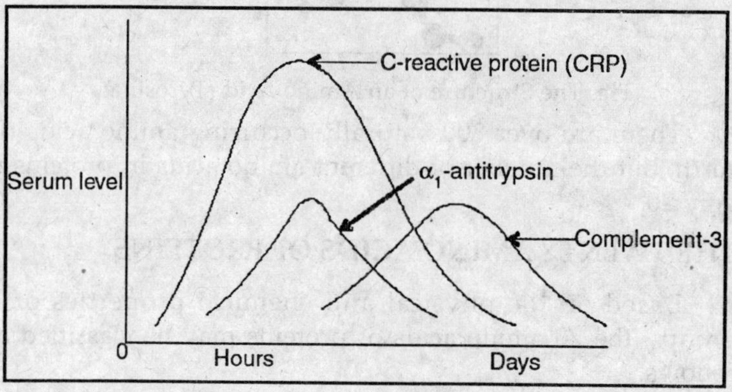

Fig. APR Levels After Injury

Globulins

The immunoglobulins and c-reactive protein (CRP) constitutes this fraction. C-reactive protein is so called because it forms precipitate with somatic C-polysaccharide of pneumococcus bacteria.

PROTEIN: STRUCTURE AND FUNCTIONS

AMINO ACIDS

An amino acid is defined as the molecule containing an

amino group (NH_2), a carboxyl group (COOH) and an R group. It has the following general formula,

$$R\text{-}CH(NH_2)\text{-}COOH$$

The R group differs among various amino acids. In a protein, the R group is also called a sidechain.

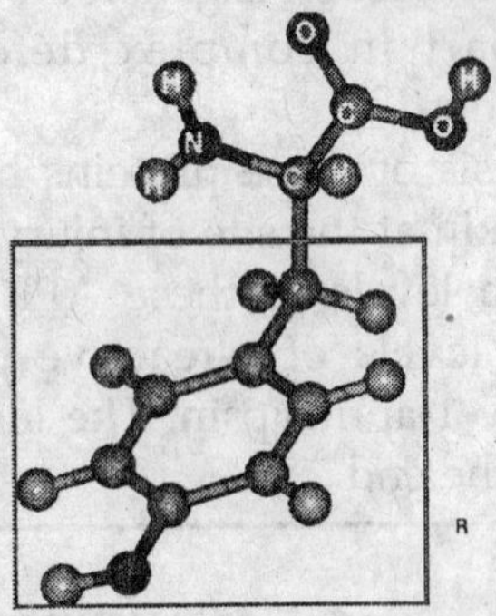

Fig. The Structure of an Amino Acid (Tyrosine).

There are over 300 naturally occurring amino acids on earth, but the number of different amino acids in proteins is only 20.

THE TWENTY AMINO ACIDS OF PROTEINS

Based on the physical and chemical properties of R groups, the 20 amino acids of proteins may be classified as follows.

- *Acidic*: Including aspartic acid (aspartate) and glutamatic acid (glutamate). In a neutral solution, the R group of an acidic amino acid may lose a proton and become negatively charged.
- *Basic*: including lysine, arginine and histidine. In a neutral solution, the R group of a basic amino acid may gain a proton and become positively charged. Interaction between positive and negative R groups may form a salt bridge, which is an important stabilizing force in proteins.
- *Aromatic*: Including tyrosine, tryptophan and phenylalanine. Their R groups contain an aromatic ring.
- *Sulfur*: Including cysteine and methionine. Their R

groups contain a sulfur atom (S). The disulfide bond formed between two cysteine residues provides a strong force for stabilizing the globular structure. A unique feature about methionine is that the synthesis of all peptide chains starts from methionine.

- *Uncharged hydrophilic*: Including serine, threonine, asparagine and glutamine. Their R groups are hydrophilic and capable of forming hydrogen bonds.

pH AND pK_a

A molecule, or an atom group in a molecule, may lose or gain a proton when the molecule is placed in an aqueous solution. The exact probability that a molecule will be protonated or deprotonated depends on the pK_a of the molecule and the pH of the solution.

Let AH be an atom group in a molecule. AH could be neutral or charged. After AH loses a proton, it is denoted by A^-. The protonation/deprotonation reaction may be written as

$$AH \rightleftharpoons H^+ + A^-$$

Define its equilibrium constant,

$$K_a = [H^+][A^-]/[AH]$$

where [..] represents the concentration.

Taking logarithm on both sides,

$$\log K_a = \log [H^+] + \log [A^-]/[AH]$$

Define

$$pK_a = -\log K_a ; \quad pH = -\log [H^+]$$

we obtain

$$pK_a = pH - \log [A^-]/[AH]$$

or

$$[A^-] / [AH] = 10^{(pH - pK_a)}$$

pH	$[H^+]$ (mol/l)	
0	10^{-0}	
1	10^{-1}	↑
2	10^{-2}	
3	10^{-3}	Increasing
4	10^{-4}	acidity
5	10^{-5}	
6	10^{-6}	
7	10^{-7}	Neutral
8	10^{-8}	
9	10^{-9}	
10	10^{-10}	Increasing
11	10^{-11}	alkalinity
12	10^{-12}	
13	10^{-13}	↓
14	10^{-14}	

From the last equation, we can see that half of molecules will lose protons if they are in a solution with pH = pK_a. The higher the pH value, the more likely a molecule will lose a proton. On the other hand, from the definition of pH, we find that the proton concentration of a solution becomes lower at higher pH, thereby capable of accepting more protons. This physical concept is consistent with the result we just obtained from the equation.

- The carboxyl group is most likely negatively charged.
- The amino group is most likely positively charged
- The R group of aspartate and glutamate are most likely negatively charged.
- The R group of lysine and arginine are most likely positively charged.
- The R group of tyrosine at pH = 7 is most likely neutral.
- The R group of histidine has 10per cent probability to become positively charged at pH = 7, but the probability increases to 50per cent at pH = 6. Thus, histidine is very sensitive to pH change in the physiological range.

Protonation/Deprotonation reaction	pK_a
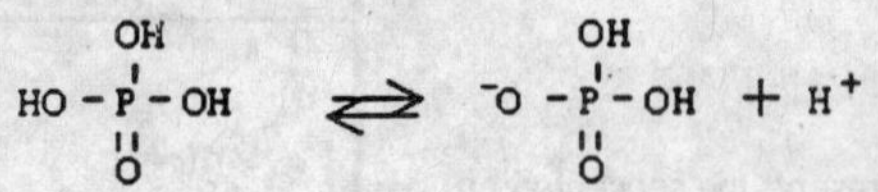	2
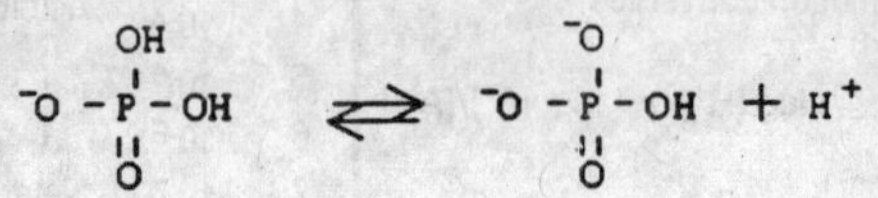	7
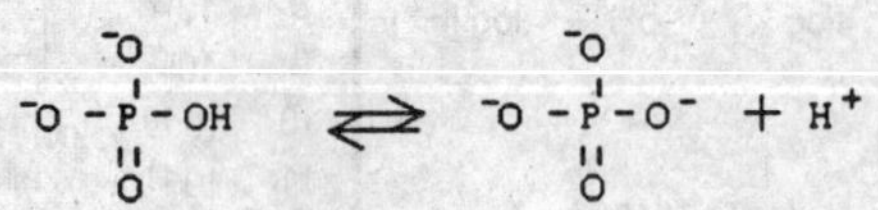	12

Fig. pK_a Values of Phosphate Groups.

This information tells us that in a neutral solution H_3PO_4 is very likely to lose a proton and become H_2PO4^-. The latter has a 50per cent chance of losing another proton, becoming HPO_4^{2-}. The chance for HPO_4^{2-} to lose an additional proton is very small. Hence, the average charge of a phosphate group is about 1.5 electron charges.

Chapter 2

Peptides

A peptide is a chain of amino acids linked together by peptide bonds. Polypeptides usually refer to long peptides whereas oligopeptides are short peptides (< 10 amino acids). Proteins are made up of one or more polypeptides with more than 50 amino acids.

Primary structure

The primary structure of a protein refers to its amino acid sequence. The amino acid in a peptide is also called a residue.

```
         1                                     20
NH3+—KETAAAKFERQHMDSSTSAASSSNY
                                             C
                                             N
                                             Q
  VAQVDALSEHVFTNVPKCRDKTLNRSKM
58C
  SQKNVACKNGQTNCYQSYSTMSITDCRETGS
                                          S 90
COO−—VSADFHVPVYPNGECAVIIHKNAQTTKYACNPYK
 124
```

Fig. The Amino Acid Sequence (Primary Structure) of Ribonuclease A (RNase A), which is an Enzyme Acting on RNA. Each Letter Represents an Amino Acid.

PEPTIDE BOND

A peptide bond is the linkage between two amino acids, formed by the condensation reaction, as shown below:

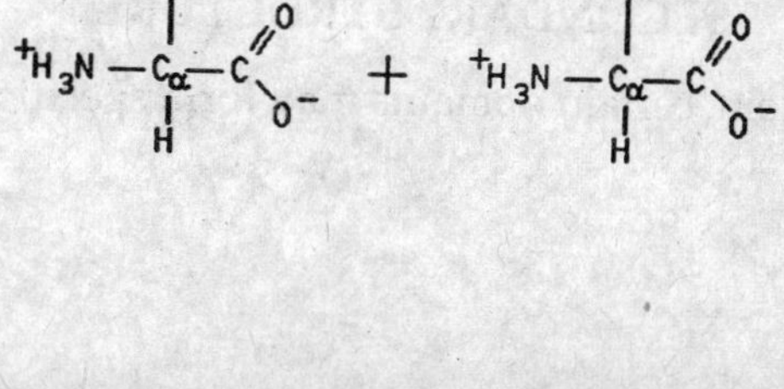

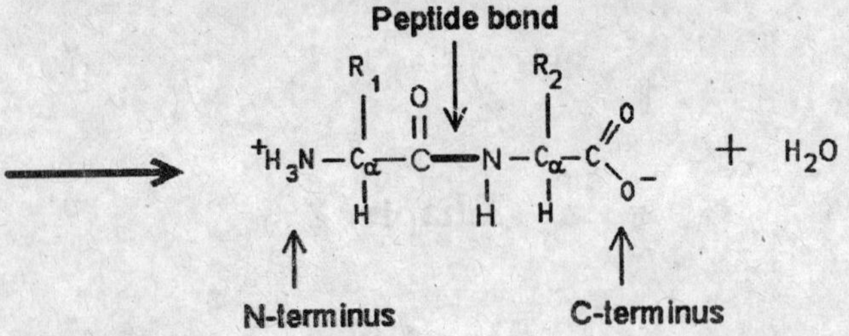

Fig. Formation of the Peptide Bond by Condensation Reaction

PHI-PSI ANGLES

Due to the specific electronic structure of the peptide bond, the atoms on its two ends cannot rotate around the bond. Hence, the atoms of the group, O=C–N–H, are fixed on the same plane, known as the peptide plane. The whole plane may rotate around the N-C_α bond (ϕ angle) or C-C_α bond (Ψ angle). C_α is the carbon atom connected to the R group.

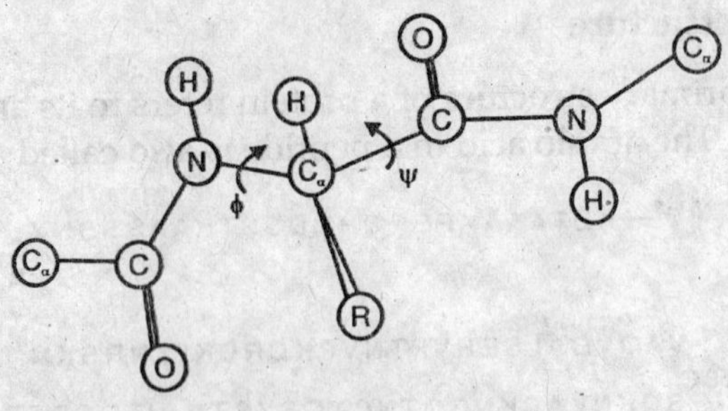

Fig. Illustration of the Peptide Plane (Gray Area) and f-Y Angles. The Red Line formed by the Repeating –C_a–C–N–C_a– is the Backbone of the Peptide Chain.

Mathematically, phi (ϕ) and psi (Ψ) are the dihedral angle (also known as torsional angle) which is defined as the angle between the point (e.g., C_α) at the end of a 4-point sequence and the plane (e.g., peptide plane) occupied by the other three points. In a peptide, phi-psi angles are restricted to certain ranges. A plot of their distribution is called the Ramachandran plot.

SECONDARY STRUCTURE

In a protein, certain domains may form specific structures

such as α helix and bstrand, which constitute the secondary structure of the protein.

```
1   KETAAAKFER QHMDSSTSAA SSSNYCNQMM KSRNLTKDRC KPVNTFVHES
       α1                    α2                    β1
51  LADVQAVCSQ KNVACKNGQT NCYQSYSTMS ITDCRETGSS KYPNCAYKTT
       α3      β2           β3          β4        β5
101 QANKHIIVAC EGNPYVPVHF DASV
           β6            β7
```

Fig. The Secondary Structure of RNase A, which Contains Three a Helices and Seven b Strands.

HYDROGEN BOND

A hydrogen bond is formed by three atoms: one hydrogen atom and two electronegative atoms (often N or O). The hydrogen atom is covalently bonded to one of electronegative atoms, called the hydrogen bond donor. The other electronegative atom is named the hydrogen bond acceptor. The two electronegative atoms may take up some electron density from the hydrogen atom. As a result, each electronegative atom carries partial negative charge and the hydrogen atom carries partial positive charge. The hydrogen atom and the hydrogen bond acceptor can then have attractive interactions.

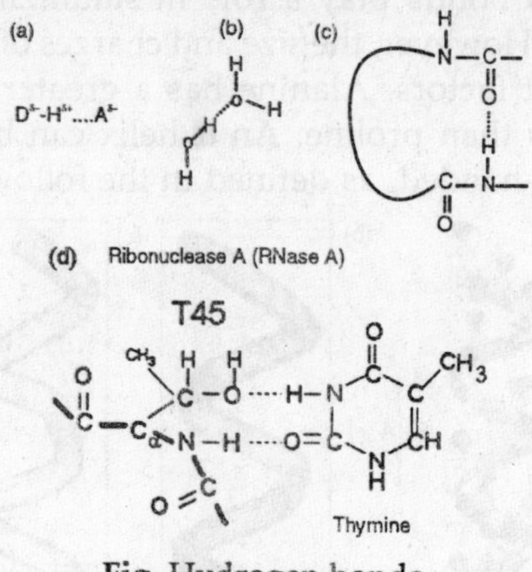

Fig. Hydrogen bonds.

(a) *The general form. D*: donor; A: acceptor.
(b) The hydrogen bond between two water molecules.
(c) The hydrogen bond in a polypeptide, e.g., in an a helix or between two b strands.

(d) The hydrogen bond between residue 45 (a threonine) of RNase A and its substrate thymine.

The strength of the hydrogen bond depends on the donor and acceptor as well as their environment. The bond energy usually ranges from 1 kcal/mol to 5 kcal/mol.

This energy is smaller than covalent bond energy, but greater than thermal energy (0.6 kcal/mol at room temperature). Therefore, the hydrogen bond can provide a significant stabilizing force in macromolecules such as proteins and nucleic acids. When a hydrogen bond is formed, the distance between the donor and the acceptor is about 0.28 nm (1 nm = 10^{-9} meter). The hydrogen atom is still covalently bonded to the donor and their distance is shorter than that between the hydrogen atom and the acceptor.

ALPHA HELIX

An α helix has the following features:

- Every 3.6 residues make one turn,
- The distance between two turns is 0.54 nm,
- The C=O (or N-H) of one turn is hydrogen bonded to N-H (or C=O) of the neighboring turn.

Hydrogen bonds play a role in stabilizing the α helix conformation. However, the size and charges of sidechains are also important factors. Alanine has a greater propensity to form α helices than proline. An α helix can be either right-handed or left-handed, as defined in the following figure.

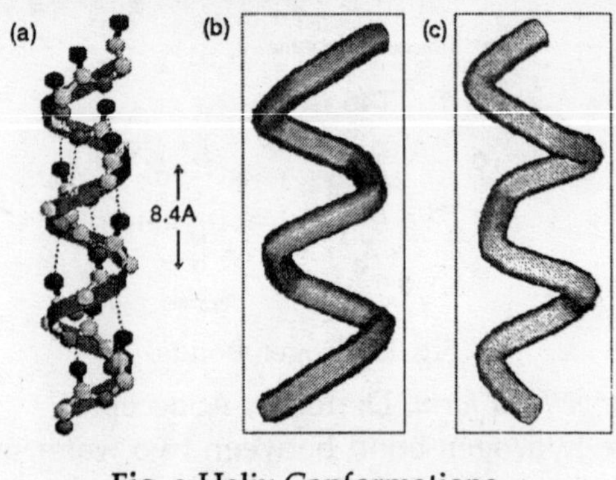

Fig. a Helix Conformations.

(a) The Ideal Right-handed a helix. C: Green; O: Red; N: Blue; H: not shown; Hydrogen Bond: dashed line.
(b) The right-handed a helix without showing atoms.
(c) The left-handed a helix.

The right-handed or left-handed helix may be distinguished by stretching out your thumb and curling the other four fingers. Imaging that in the above figure the helix is spiraling upward. Stretch your right thumb upward, then the other four fingers of the right hand will be able to curl in the same direction as each turn in the spiral. The same result can be obtained if you imagine that the helix is spiraling downward and point your thumb downward. The left hand works for the left-handed a helix.

AMPHIPATHIC ALPHA HELIX

In an amphipathic α helix, one side of the helix contains mainly hydrophilic amino acids and the other side contains mainly hydrophobic amino acids. The amino acid sequence of amphipathic α helix alternates between hydrophilic and hydrophobic residues every 3 to 4 residues, since the α helix makes a turn for every 3.6 residues. An example is shown below.

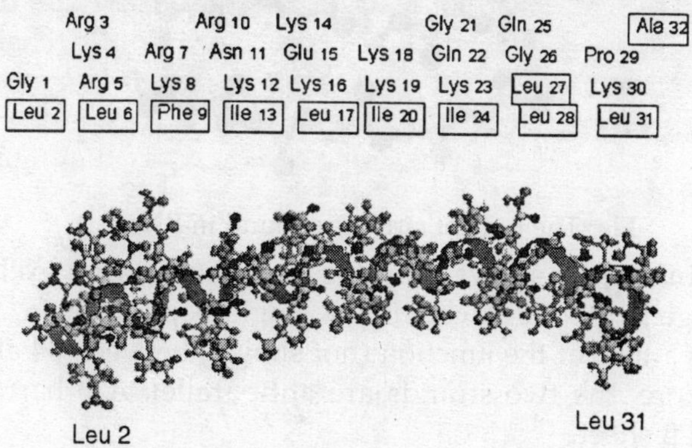

Fig. The amphipathic a helix structure of CAP18, which is a molecule capable of binding to the endotoxin of bacteria.

(a) Amino acid sequence of the amphipathic part of CAP18. Hydrophobic residues are boxed with red lines.

(b) The 3D structure determined by nuclear magnetic resonance. Hydrophobic residues are located on the lower side. PDB ID = 1LYP.

BETA STRAND, BETA SHEET AND BETA BARREL

Beta Strand

In a β strand, the torsion angle of N-C_α-C-N in the backbone is about 120 degrees. The following figure shows the conformation of an ideal β strand. Note that the sidechains of two neighboring residues project in the opposite direction from the backbone.

Beta Sheet

A β sheet consists of two or more hydrogen bonded β strands. The two neighboring b strands may be parallel if they are aligned in the same direction from one terminus (N or C) to the other, or anti-parallel if they are aligned in the opposite direction.

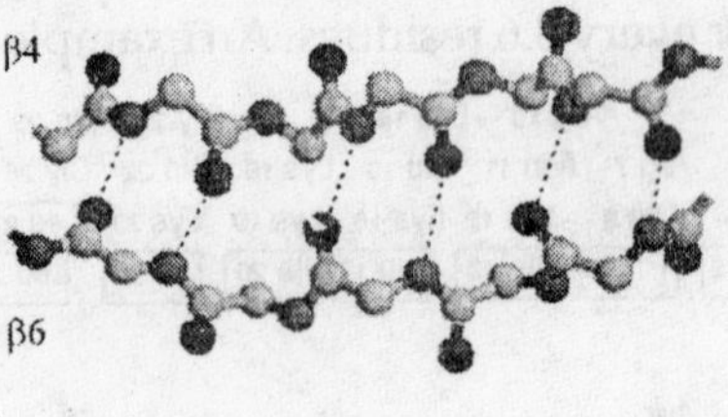

Fig. The β sheet structure found in RNase A.

This figure shows only the backbone atoms, excluding hydrogens. RNase A contains a single peptide chain, which makes a turn at the junction (not shown) between β4 and β6. Therefore, the two strands are antiparallel. A β barrel is a closed β sheet.

PROTEIN MOTIFS AND DOMAINS

The motif is a characteristic domain structure consisting of two or more a helices or b strands. Common examples include coiled coil, helix-loop-helix, zinc finger, leucine zipper, etc.

Many proteins also contain specific domains such as the SH2 domain.

HELIX-LOOP-HELIX

"Helix-Loop-Helix" is characterized by two a helices connected by a loop. It should not be confused with "Helix-Turn-Helix".

In fact, these two structures are quite different, but the Helix-Loop-Helix looks like leucine zipper. Example: myoblast determination proteins (MyoD).

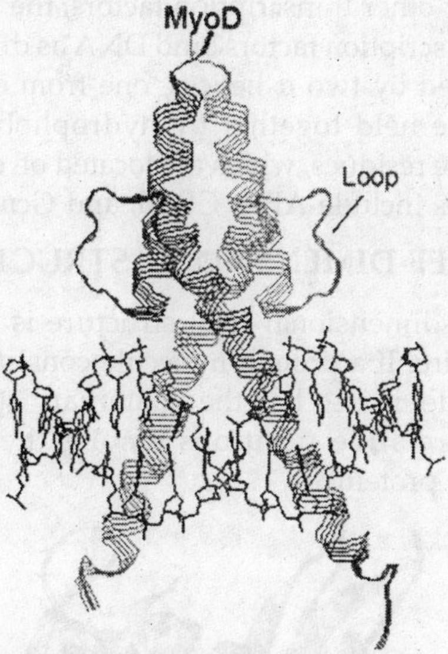

Fig. The transcription factor MyoD.

(a) The domain organization.

(b) The structure of the helix-loop-helix motif. PDB ID = 1MDY.

Zinc Finger

The zinc finger motif contains one or more zinc ions which are crucial for the structural stability. It can be divided into three types: C_2H_2 zinc finger: It is characterized by the sequence $CX_{2-4}C....HX_{2-4}H$, where C = cysteine, H = histidine,

X = any amino acid. In the 3D structure, two cysteine residues and two histidine residues interact with a zinc ion.

C_4 zinc finger: Its consensus sequence is $CX_2CX_{13}CX_2CX_{14-15}CX_5CX_9CX_2C$. The first four cysteine residues bind to a zinc ion and the last four cysteine residues bind to another zinc ion. C_6 zinc finger. It has the consensus sequence $CX_2CX_6CX_{5-6}CX_2CX_6C$. The yeast's Gal4 contains such a motif where six cysteine residues interact with two zinc ions.

Leucine Zipper

Like many other transcription factors, the leucine-zipper-containing transcription factors bind DNA as dimers. A leucine zipper is formed by two α helices, one from each monomer. The helices are held together by hydrophobic interactions between leucine residues, which are located on one side of each helix. Examples include AP-1, CREB, and Gcn4.

THREE-DIMENSIONAL STRUCTURE

The three-dimensional (3D) structure is also called the tertiary structure. If a protein molecule consists of more than one polypeptide, it also has the quaternary structure, which specifies the relative positions among the polypeptides (subunits) in a protein.

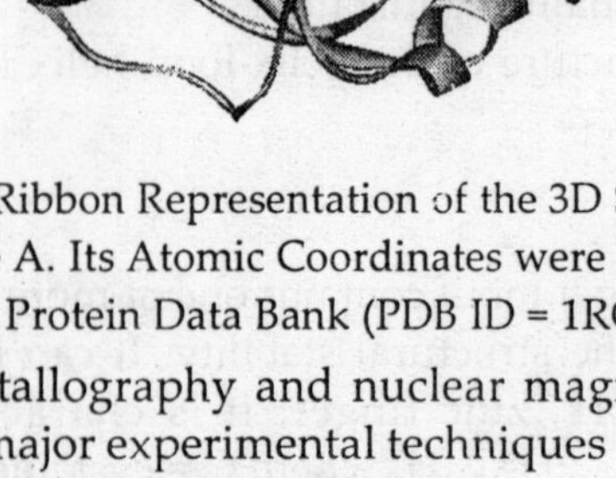

Fig. The Ribbon Representation of the 3D Structure of RNase A. Its Atomic Coordinates were obtained from Protein Data Bank (PDB ID = 1RCN).

X-ray crystallography and nuclear magnetic resonance (NMR) are the major experimental techniques for determining the 3D structures of macromolecules.

Protein Data Bank is the most important site for structural information which can easily be accessed by the "PDB ID" of a macromolecule.

ENZYMES

Enzymes are the catalysts of biochemical reactions in the cell. Numerous drugs are enzyme inhibitors.

Prominent enzymes as drug targets:

- *HIV Protease*: Essential for HIV replication. Its inhibitor has been able to control AIDS.
- *Angiotensin Converting Enzyme (ACE)*: Promotes contraction of blood vessels. Its inhibitors have been widely used to treat high blood pressure and congestive heart failure.
- *HMG-CoA Reductase*: Essential for the synthesis of cholesterol. Its inhibitors (such as statins) can reduce the cholesterol level.
- *cGMP Phosphodiesterase*: Catalyzes the conversion of cGMP into GMP. It is the target of the well known drug for treating impotence, Viagra.
- *Calcineurin*: A phosphatase that plays an important role in immune responses. It is the target of the immunosuppressive drugs FK506 and cyclosporin A.

ENZYME INHIBITORS

Numerous drugs are enzyme inhibitors which may be divided into two types: competitive inhibitors and non-competitive inhibitors.

The competitive inhibitor directly occupy the active site before the substrate can bind to it.

The non-competitive inhibitor does not occupy the active site, but its binding to other region somehow inhibits the catalytic activity.

HIV protease inhibitor is a competitive inhibitor.

Angiotensin II

We are talking about the chemical messengers in the body

called neurohormones. For those of us with heart failure, the neurohormone called angiotensin II is not a good thing.

Angiotensin II helps controls the cardiovascular system. It causes blood vessels to shrink (vasoconstriction), which raises blood pressure. Angiotensin II also causes the body to release aldosterone - a substance that causes our kidneys to retain sodium and fluid, causing edema. Angiotensin II also causes the body to release a substance called vasopressin or ADH, which causes us to retain fluid even more. Angiotensin II is a real bug-a-boo for us CHFers.

ACE Inhibitors

We all have angiotensin one in our systems. Angiotensin one is converted into angiotensin II by an enzyme in your body called ACE (Angiotensin Converting Enzyme). ACE is what we block with ACE inhibitors - blocking it prevents angiotensin I from converting into angiotensin II.

Reducing angiotensin II lets our blood vessels relax and expand. This lowers blood pressure, which makes the heart's job a lot easier.

Think of it this way: If you pump the same amount of water at the same speed through a little pipe as through a big pipe, there's higher pressure in the small one. The fluid *shoots* out of the small pipe, but calmly flows out of the big one, even though they are both moving the same amount of water. The pump pushing the water through that skinny pipe has to work a lot harder than the pump glushing it through the big pipe.

The same principle applies to our blood vessels. If our arteries relax, they get bigger (vasodilation) and the pump has an easier job - less pressure. That's called reducing "afterload." ACE inhibitors do this for us. "Super" ACE inhibitors called vasopeptidase inhibitors were also tested for CHF but carry higher risk for side effects.

ACE also helps control the amount of blood in your body - your blood "volume." Blocking ACE reduces the overall amount of blood in your body. This means your tired old heart has less blood to pump, which makes its job easier. ACE inhibitors increase the level of potassium in your blood, so they partly offset potassium loss caused by diuretics like Lasix and Demadex.

Finally, ACE inhibitors help us live longer and feel better. The articles below show that ACE inhibitors reduce our risk of death by 20per cent to 40per cent. The other two types of drugs that do this are beta-blockers and Aldactone (spironolactone). ACE inhibitors slow heart remodeling, preventing our hearts from getting even weaker over time.

It can take several months for the full effect of ACE inhibitors to show up, and improve our quality of life. Hang in there, because ACE inhibitors *do* work and *all* CHFers should be on them if possible. It's even in the official heart failure treatment guidelines.

Unfortunately, many doctors don't offer us ACE inhibitors if we have any kidney problems or if our blood pressure is low. Monopril (fosinopril) is processed by the liver as well as the kidneys, so CHFers with kidney problems can usually take it okay. People with low blood pressure, surprisingly enough, often tolerate ACE inhibitors quite well (although not always).

Angiotensin II Receptor Blockers[ARBs]

Angiotensin II is also made at other places in the body. So ACE inhibitors can't *completely* stop angiotensin II from affecting you. After starting an ACE inhibitor, there is a period where angiotensin II is almost completely stopped. Over time, your angiotensin II level rises again, from these other places in the body.

Since ACE inhibitors can't do it all, it seems like a good idea to block the stuff at other places too. ARBs block angiotensin II receptors on cell walls. If angiotensin II can't connect to cell receptors, it can't affect us. The most important receptor is called the AT-1 receptor (there is also an AT-2 receptor). ARBs *do* help CHF. The question is: Do ARBs help CHFers already taking an ACE inhibitor? This has not been completely decided yet. The resolved study compared an ARB alone to using an ACE inhibitor alone to using both an ACE inhibitor and an ARB. Using both worked better to improve heart enlargement and EF.

The ValHeft trial added an ARB to standard CHF therapy. Complications and mortality improved most in CHFers taking the ARB with either an ACE inhibitor *or* a beta-blocker, but not in those taking all 3 drugs. Why taking all 3 drugs gave worse results is unknown.

ARBs do have some excellent uses we *are* sure about. In patients who develop a bad cough from using an ACE inhibitor, an ARB can often be used without the cough. In patients who don't tolerate an ACE inhibitor for some other reason, an ARB will often do the job.

High doses are better than low doses but doctors don't usually prescribe them - due to simple ignorance. Double-check your doc on this to be sure you are getting a high enough dose!

Even if you improve to normal heart function and consider yourself "cured," a good CHF specialist will keep you on an ACE inhibitor for life, and that's smart. It may stop you from getting CHF again down the road - which can happen.

ACE inhibitor use may reduce your zinc level so you might want to take a zinc supplement. Your sense of taste could possibly go goofy on you after starting an starting ACE inhibitor too - especially on captopril.

Synthesis of Lipids

Lipid synthesis occurs at the smooth endoplasmic reticulum. Lipids may be transported to appropriate locations by vesicles or by phospholipid exchange proteins.

Fueled by fat-rich diets and sedentary lifestyles, atherosclerosis and obesity are major health issues in the Western world. Both diseases involve excessive accumulation of neutral lipids: cholesterol esters in arteries in atherosclerosis and triglycerides in adipocytes in obesity. The study of these diseases has sparked interest in the biochemistry and molecular biology of neutral lipid synthesis.

The discovery of acyl-CoA:cholesterol acyltransferase (ACAT,[1] EC 2.3.1.26), a cholesterol ester synthesis enzyme, dates back to the 1950s. ACAT was identified as an intracellular enzyme in the endoplasmic reticulum (ER) that covalently joined cholesterol and fatty acyi-CoA molecules to form cholesterol esters. ACAT activity was detected in macrophages and in tissues such as liver, small intestine, and adrenal glands. Another ER enzyme, acyl-CoA:diacylglycerol acyltransferase (DGAT, EC 2.3.1.20), was identified in 1960. The DGAT reaction is similar to that of ACAT except that diacylglycerol is the acyl group acceptor. In the ensuing 30 years, much was learned about the biochemistry of ACAT and DGAT.

Both enzymes were found to play important roles in synthesizing neutral lipids for the assembly and secretion of lipoproteins, and ACAT was found to be responsible for synthesizing cholesterol esters in arterial macrophage foam cells. However, these hydrophobic proteins proved difficult to isolate, slowing progress in their understanding.

Because of the pioneering work of Chang and colleagues the field of neutral lipid synthesis made great strides during the past decade. These investigators isolated ACAT by using a clever cloning strategy to isolate an ACAT cDNA. Thanks to the resultant molecular probes and to genetic studies in yeast and mice, we now know there are two mammalian ACAT enzymes and probably more than one DGAT enzyme.

A large number of ACAT- and DGAT-related genes are identifiable in species ranging from plants to yeast to humans. Here we review recent advances in understanding the function of the mammalian members of this family, ACAT1, ACAT2, and DGAT.

Gene and Protein Structures

The human ACAT1 gene encodes four mRNAs of 7.0, 4.3, 3.6, and 2.8 kilobases (kb) (12, 18-20); all contain the same translational reading frame but differ in the length of the untranslated regions. The two shorter mRNAs are products of a proximal ACAT promoter.

The 4.3-kb mRNA is derived from an unusual RNA recombination mechanism involving *trans*-splicing of two discontinuous precursor RNAs produced from chromosomes 1 and 7 (21). The origin of the 7.0-kb transcript is unknown. The human ACAT2 gene encodes a single mRNA of 2.2 kb, and the human DGAT gene encodes two mRNAs of 2.0 and ?.4 kb. Regions of shared sequence identity are present throughout the ACAT and DGAT proteins, but the greatest similarity is found in their C termini. Human and mouse ACAT2 are 40per cent identical to human ACAT1, and DGAT is ~20per cent identical to ACAT1.

Little is known about the relationship between protein structure and function in ACAT/DGAT family members. One motif, FYXDWWN (amino acids 403-409 of human ACAT1), is highly conserved in all family members and may be involved in fatty acyl-CoA binding. Another motif, MKXXSF (amino acids 265-270 of human ACAT1), is conserved in ACAT family members.

The serine of this motif is required for ACAT activity. All three family members have a potential tyrosine phosphorylation motif and at least one *N*-linked glycosylation site, both of unknown significance. ACAT1 and ACAT2 contain an N-terminal leucine zipper motif. Although ACAT1 functions as a homotetramer it is unknown whether this leucine zipper motif participates in tetramer formation or whether ACAT2 and DGAT act as multimers. ACAT1 and ACAT2 do not form hetero-oligomeric complexes.

All ACAT/DGAT proteins possess multiple hydrophobic regions predicted to serve as transmembrane domains, although there is not yet a consensus structural model for the proteins. Investigations of human ACAT1 topology have yielded different results. One study, performed by inserting epitopes at different positions in the expressed protein found seven transmembrane domains.

Another study, performed by expressing truncated forms of the protein found five transmembrane domains. In both studies, the N terminus of the protein localized to the cytoplasm and the C terminus to the ER lumen. Topologic studies of human ACAT2 performed with the truncation mutant method also found five transmembrane domains with the N terminus in the cytoplasm and the C terminus in the ER lumen.

The latter study found that a serine residue essential for

ACAT activity was located in the cytoplasm for ACAT1 and in the ER lumen for ACAT2, suggesting that the active sites of these enzymes reside on opposite sides of the ER membrane. Several putative transmembrane sequences are conserved in ACAT1 and ACAT2 but not in DGAT suggesting that these regions are involved in cholesterol binding. The topology of DGAT has not been reported, but a putative diacylglycerol binding motif has been identified.

TISSUE AND CELLULAR DISTRIBUTIONS

The tissue distributions of ACAT1 and ACAT2 are largely complementary. ACAT1 mRNA is present in many tissues, with the highest levels in macrophages and in adrenal and sebaceous glands all of which store cholesterol esters in cytoplasmic droplets.

ACAT1 is also highly expressed in human atherosclerotic lesions, particularly in macrophage foam cells. ACAT2 is expressed predominantly in the liver and small intestine. In humans, nonhuman primates, and mice, ACAT2 appears to be the major ACAT in the small intestine and the predominant isozyme in hepatocytes of nonhuman primates and mice.

Whether ACAT1 or ACAT2 is the predominant enzyme in human hepatocytes is controversial. Both are expressed in human hepatocytes and HepG2 cells. However, immunodepletion experiments found that ACAT1 accounts for ~90per cent of ACAT activity in human liver, whereas ACAT2 accounts for most of the ACAT activity in adult small intestine and fetal liver but only 10-20per cent of activity in adult liver.

It is unclear whether the hepatic ACAT1 activity resides in hepatocytes or macrophage-derived Kupffer cells. ACAT1 expression was detected immunohistochemically in both cell types, with stronger staining in Kupffer cells. In nonhuman primates, ACAT1 was expressed only in Kupffer cells in the liver. A significant proportion of the ACAT1 activity in human liver thus may be attributable to Kupffer cells. Another study found little evidence of ACAT2 expression in human hepatocytes.

DGAT mRNA and activity are ubiquitous in mouse and

human tissues, with the highest levels in liver, small intestine, and adipose tissue and somewhat lower levels in testis and adrenal gland. This wide range of tissue expression is consistent with the involvement of DGAT in the glycerol phosphate pathway of triglyceride synthesis, which is common to most cells. DGAT protein expression in tissues has not been studied because of the lack of suitable antibodies.

The intracellular localization of enzymes in this class has been examined only for ACAT1. In human melanoma cells and fibroblasts and in mouse macrophages, ACAT1 is located primarily in the ER. In mouse macrophages, a small portion of ACAT1 immunoreactivity localizes to a region near the *trans*-Golgi network. In macrophages, ACAT1 localization may change with different conditions.

BIOCHEMICAL ACTIVITY AND REGULATION

ACAT and DGAT substrates have been identified by expressing the proteins in insect cells or yeast devoid of sterol esterification activity. In insect cells, mouse and human ACAT1 and ACAT2 utilize a variety of oxysterols in addition to cholesterol as substrates. Ergosterol and plant sterols are poor substrates.

Both enzymes also utilize a wide variety of long chain fatty acyl-CoAs as substrates although arachidonyl-CoA appears to be a better substrate for ACAT1 than ACAT2. In insect cells, DGAT utilizes only diacylglycerol as the acyl acceptor. DGAT has a broad fatty acyl-CoA specificity.

ACAT1 is regulated primarily by post-translational mechanisms. In cholesterol-loaded cells, esterification increases without changes in the expression level of ACAT1 mRNA or protein. Purified ACAT1 is activated allosterically by cholesterol or oxysterols. ACAT1 activity may also be regulated by its intracellular localization. For example, a portion of ACAT1 immunoreactivity in macrophages is found near the *trans*-Golgi network and the endocytic recycling compartment.

These cholesterol-rich organelles are involved in membrane traffic with the plasma membrane, and the proximity of ACAT1 to these organelles may be important for cholesterol ester synthesis during the formation of macrophage foam cells. The observation that 30-40per cent of ACAT1 redistributed from the ER to small vesicles during foam cell formation is consistent with this possibility. Regulation by phosphorylation has not been demonstrated.

ACAT1 is also transcriptionally regulated; however, only the expression levels of the shorter ACAT1 mRNAs (2.8 and 3.6 kb) change. In mice and rabbits a fat and cholesterol-rich diet increases ACAT1 mRNA levels 2-3-fold in the liver. In human HepG2 cells, free fatty acids, but not 25-hydroxycholesterol, increase ACAT1 mRNA levels 1.5-2-fold. ACAT1 mRNA and protein expression levels also increase during the differentiation of monocytes into macrophages.

Two factors that induce gene expression during monocyte-macrophage differentiation1,25-dihydroxyvitamin D_3 and 9-*cis*-retinoic acid, also increased ACAT1 mRNA, protein, and cholesterol esterification activity in THP-1 cells.

Little is known about the regulation of ACAT2 and DGAT. Levels of DGAT mRNA increase markedly in parallel with enzyme activity in NIH 3T3-L1 cells during their differentiation into adipocytes.

This finding suggests that transcription factors involved in adipogenesis activate DGAT expression. Like other family members, DGAT possesses a putative tyrosine phosphorylation site. The use of this site in DGAT regulation has not been demonstrated, although two reports have indicated that the enzyme may be posttranslationally regulated by a tyrosine kinase.

FUNCTIONAL ANALYSIS OF ACAT1 AND ACAT2

Gene knockout experiments in mice have provided insights into the biological roles of ACAT1 and ACAT2. In ACAT1-deficient mice, plasma cholesterol levels, hepatic cholesterol esterification activity, and intestinal cholesterol absorption are normal; consistent with the presence of a second ACAT enzyme in the small intestine and liver of mice.

ACAT1 deficiency has its greatest effect in the adrenal

glands and macrophages. Cholesterol esters and cholesterol esterification activity are virtually undetectable in the adrenal cortex of ACAT1$^{-/-}$ mice, although the depletion of cholesterol esters does not affect corticosterone synthesis. In addition, peritoneal macrophages from ACAT1$^{-/-}$mice do not accumulate cholesterol esters when incubated with cholesterol-containing lipoproteins.

This finding provides a unique tool for testing the role of ACAT1 in macrophage foam cell formation and atherosclerosis.

Another ACAT1 gene disruption study in mice confirmed general aspects of the ACAT1$^{-/-}$ phenotype and showed that ACAT1 deficiency reduces cholesterol esterification activity by >95per cent in testis and ovary and results in atrophy of the meibomian glands of the eyelids. A naturally occurring mutation in the mouse ACAT1 gene is present in the AKR strain of inbred mice which are homozygous for the *adrenocortical lipid depletion* (*ald*) allele.

The analysis of ACAT2-deficient mice (ACAT2$^{-/-}$) complemented findings from ACAT1$^{-/-}$ mice. ACAT2$^{-/-}$ mice have nearly undetectable ACAT activity and lack cholesterol esters in the liver and small intestine. Because they cannot synthesize cholesterol esters in the small intestine, ACAT2$^{-/-}$ mice have a reduced capacity to obtain cholesterol from the diet. As a result, they do not develop hypercholesterolemia or form cholesterol gallstones when fed a diet high in fat, cholesterol, and cholic acid.

ACAT2 deficiency markedly decreases the cholesterol ester content of apolipoprotein (apo) B-containing lipoproteins in mice fed either a chow diet or the above diet. Thus, ACAT2 is the cholesterol esterification enzyme in mouse liver and small intestine and plays a major role in mediating the response to dietary cholesterol. Gene disruption studies suggest that both ACAT enzymes synthesize cholesterol esters: ACAT1 for storage in cytosolic droplets and ACAT2 for secretion in apoB-containing lipoproteins. However, these results may reflect their different tissue distributions rather than inherent biochemical differences.

Both ACAT1 and ACAT2, when expressed in a cell line deficient in cholesterol esterification activity (AC29 cells), synthesize cholesterol esters that are stored in cytoplasmic droplets (25). Also, overexpression of ACAT1 in mouse liver causes both cholesterol ester storage and secretion.

ROLE OF ACAT1 IN ATHEROSCLEROSIS

Because cholesterol ester synthesis by ACAT1 is involved in macrophage foam cell formation, it has been hypothesized that inhibition of ACAT1 may reduce atherosclerosis. To test the hypothesis that ACAT1 deficiency in macrophages inhibits atherosclerosis, ACAT1$^{-/-}$ mice were crossed with two strains of mice that are susceptible to atherosclerosis: low density lipoprotein receptor-deficient (LDLR$^{-/-}$) mice and apoE-deficient (apoE$^{-/-}$) mice.

ACAT1 deficiency did not prevent atherosclerotic lesions in either model. Because ACAT1 deficiency also reduced serum cholesterol levels in both strains of mice, its effect on lesion size was confounded by effects on serum cholesterol levels, and thus the contribution of ACAT1 to plaque formation could not be rigorously determined. Qualitative analysis of lesions in the mutant mice, however, revealed reduced amounts of cholesterol esters and macrophages. These alterations may increase lesion stability although this hypothesis remains to be tested.

Studies that analyzed the quantitative effect of ACAT1 deficiency on lesion development have produced different results. Ishibashi and colleagues crossed a different strain of ACAT1$^{-/-}$ mice with LDLR$^{-/-}$ and apoE$^{-/-}$ mice and found that ACAT1 deficiency was associated with ~50per cent reduction in aortic root lesions in LDLR$^{-/-}$ mice and did not decrease serum cholesterol levels.

However, Fazio *et al.* showed that LDLR$^{/}$mice reconstituted with ACAT1$^{-/-}$ macrophages had increased atherosclerotic lesions that contained more free cholesterol and fewer macrophages than control lesions. The apparent discrepancies between these studies may reflect differences in methodology or study design.

Ishibashi and colleagues quantified lesion areas by measuring neutral lipid staining even though the cholesterol ester content of atherosclerotic lesions in ACAT1$^{-/-}$ mice was significantly diminished.

The lesion areas consequently may have been underestimated. In the study by Fazio *et al*, selective inhibition of macrophage ACAT1 may not have reduced lesion size because ACAT activity was not inhibited in adjacent arterial smooth muscle cells, which can also become foam cells.

Both complete and macrophage-specific ACAT1 deficiencies have deleterious effects in hypercholesterolemic mice. ACAT1 deficiency causes extensive deposition of free cholesterol in the skin and brain of LDLR$^{-/-}$ and apoE$^{-/-}$ mice, suggesting an important role for the enzyme in cholesterol metabolism in these tissues when serum sterol levels are elevated.

Although the dramatic effects in these models reflect the extreme condition of complete ACAT1 inhibition in the face of severe hypercholesterolemia, they indicate that selective ACAT1 inhibition can be detrimental.

FUNCTIONAL ANALYSIS OF DGAT

Because DGAT catalyzes the final step in the major pathway of triglyceride synthesis it was speculated that inactivation of the encoding gene in the mouse (*Dgat*) would cause severe morbidity or early lethality.

Surprisingly, DGAT-deficient mice are viable and appear healthy although adult mice develop dry fur and female mice have a lactation defect. Even though these animals lack triglyceride synthesis in most tissues examined, *Dgat*$^{-/-}$ mice have normal serum triglyceride levels and triglycerides in white adipose tissue. These results indicate the existence of alternative pathways of triglyceride synthesis.

Although DGAT deficiency does not prevent triglyceride synthesis, *Dgat*$^{-/-}$ mice have~50per cent reduction in fat pad content and are resistant to weight gain when fed a high fat diet. Unexpectedly, these findings are associated with an ~15per cent increase in daily total energy expenditure, part of which is mediated by increased physical activity when the mice are fed a high fat diet.

The mechanism for the increased energy expenditure is unknown, but the *Dgat*$^{-/-}$ phenotype indicates that the disruption of triglyceride synthesis significantly affects energy metabolism.

ACAT AND DGAT ENZYMES AS PHARMACEUTICAL TARGETS

Because ACAT enzymes participate in cholesterol ester synthesis for assembly of atherogenic apoB-containing lipoproteins and in macrophage foam cell formation, ACAT inhibitors have been developed as potential therapeutics. These agents produced mixed results in several animal models.

In cholesterol-fed rabbits and hamsters, ACAT inhibitors reduced atherosclerotic lesion development independently of their effects on serum cholesterol levels.

A study of genetic models of hypercholesterolemia in mice and rabbits suggested that ACAT inhibitors either had no effect or increased aortic sinus lesion size.

One inhibitor (avasimibe) reduced atherosclerosis in several animal models and is being evaluated in clinical trials. This compound, like most currently available ACAT inhibitors, is not selective for one of the two ACAT enzymes.

This pan-specificity may be desirable because pharmacologic and genetic studies in animals suggest that specific inhibition of macrophage ACAT1 may increase lesion size, possibly because of toxicity from excess free cholesterol. Selective inactivation of ACAT1 in hypercholesterolemia may also have detrimental consequences.

Therefore, selective ACAT1 inhibition should be approached cautiously in humans. Partial inhibition of both ACAT enzymes may be a preferable strategy. A selective inhibitor of ACAT2 may be useful for preventing diet-induced hypercholesterolemia.

Recent findings in *Dgat*$^{-/-}$ mice 44447have sparked interest in DGAT inhibitors as possible therapeutic agents for obesity. Several naturally occurring compounds have been identified.

Signaling Via Hydrophobic Molecules

Hydrophobic molecules can move in and out of cells by passing through lipid bilayers. Nitric oxide, arachidonic acid and steroids have been shown to play important ıoles in cell signaling.

NITRIC OXIDE

Nitric oxide (NO) is produced from the following reaction:

$$\text{L-arginine} \xrightarrow{\text{NOS}} \text{L-citrulline} + \text{NO}$$

Where NOS represents NO synthase. After NO is generated in a cell, it diffuses to act on target molecules in neighboring cells.

NO may stimulate soluble guanylyl cyclase to produce cGMP which regulates several enzymes and ion channels. In smooth muscle, an important action of cGMP is to induce muscle relaxation. Normally, cGMP will soon be converted into GMP by phosphodiesterase. The well known drug for impotence, Viagra (sildenafil citrate), inhibits phosphodiesterase.

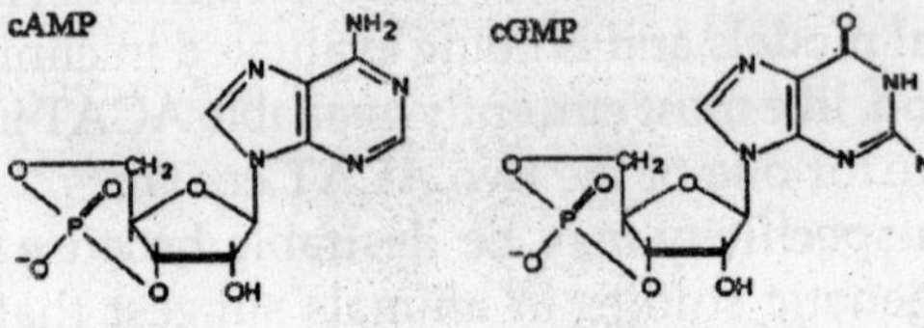

Fig. Structures of cAMP and cGMP

Arachidonic Acid

Arachidonic acid is generated from phospholipid hydrolysis catalyzed by phospholipase. Like NO, its target is usually not located in the cell where it was first generated. After diffusing to target cells, arachidonic acid may activate protein kinase, resulting in phosphorylation of target molecules. Many of its target molecules are involved in learning and other neuronal activities.

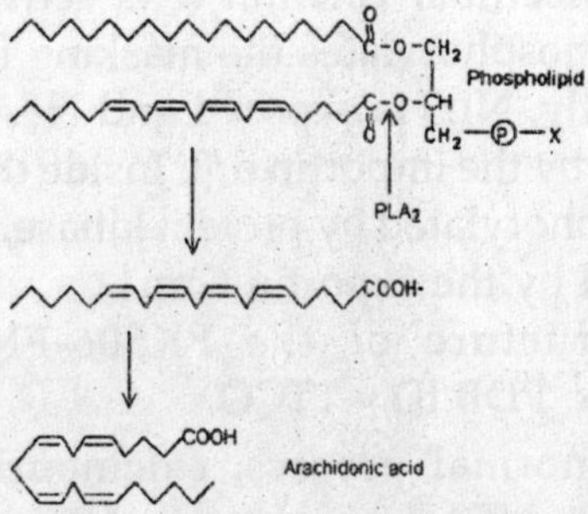

Fig. Generation of Arachidonic Acid from Phospholipid by Phospholipase A_2 (PLA_2).

Steroids

The major role of steroids is to regulate transcription, since many steroid receptors are transcription factors.

NFAT, Calcineurin and Immunosuppression

Organ transplantation often elicits immune responses to reject the foreign object. The most widely used drugs to deal with such graft reaction are FK506 and cyclosporin A (CsA). Their direct target is a phosphatase known as calcineurin. CsA combines with a cellular protein called cyclophilin to inhibit calcineurin; FK506 combines with FKBP (FK506 binding protein) to inhibit the phosphatase.

How could inhibition of a phosphatase suppress the immune response? The answer lies in the transport of NFAT (nuclear factor of activated T cells) which is a transcription factor regulating IL-2.

NFAT contains both nuclear localization sequence (NLS) and nuclear export sequence (NES), but one of them is buried in the protein interior, inaccessible to importin and exportin. Whether NLS or NES is masked depends on the phosphorylation state of specific serine residues in the regulatory domain.

Phosphorylation of these serine residues exposes NES whereas dephosphorylation exposes NLS. In resting cells, the NLS of the cytoplasmic NFAT is masked due to phosphorylation on these serine residues. In stimulated cells, an increase of intracellular calcium ions activates calcineurin, which then dephosphorylates the masking residues.

Consequently, NLS is exposed and NFAT can be carried into the nucleus by the importin α/β. Inside the nucleus, NFAT may be re-phosphorylated by protein kinase, exposing its NES and be exported by the exportin Crm1.

- The structure of the FK506-FKBP-calcineurin complex. PDB ID = 1TCO.
- In the normal process, calcineurin may dephosphorylate NFAT, exposing its NLS so that NFAT can move into the nucleus to stimulate the expression of IL-2, inducing a cascade of immune responses. If calcineurin is inhibited, the immune response will be suppressed.

CLASSIFICATION OF ENZYMES

Based on catalyzed reactions, the nomenclature committee of the International Union of Biochemistry and Molecular Biology (IUBMB) recommended the following classification.

- Oxidoreductases catalyze a variety of oxidation-reduction reactions. Common names include dehydrogenase, oxidase, reductase and catalase.
- Transferases catalyze transfers of groups (acetyl, methyl, phosphate, etc.). Common names include acetyltransferase, methylase, protein kinase and polymerase. The first three subclasses play major roles in the regulation of cellular processes. The polymerase is essential for the synthesis of DNA and RNA.
- Hydrolases catalyze hydrolysis reactions where a molecule is split into two or more smaller molecules by the addition of water. Common examples are given below.

Proteases splits protein molecules. Examples: HIV protease and caspase. HIV protease is essential for HIV replication. Caspase plays a major role in apoptosis.

Apoptosis

Apoptosis is a tightly regulated form of cell death, also called the programmed cell death. Morphologically, it is characterized by chromatin condensation and cell shrinkage in the early stage.

Then the nucleus and cytoplasm fragment, forming membrane-bound apoptotic bodies which can be engulfed by phagocytes. In contrast, cells undergo another form of cell death, necrosis, swell and rupture. The released intracellular contents can damage surrounding cells and often cause inflammation. Apoptosis is an important process during normal development. It is also involved in aging and various diseases such as cancer, AIDS, Alzheimer's disease and Parkinson's disease.

THE MECHANISM

During apoptosis, the cell is killed by a class of proteases called caspases. More than 10 caspases have been identified. Some of them (e.g., caspase 8 and 10) are involved in the initiation of apoptosis, others (caspase 3, 6, and 7) execute the death order by destroying essential proteins in the cell. The apoptotic process can be summarized as follows:

- Activation of initiating caspases by specific signals.
- Activation of executing caspases by the initiating caspases which can cleave inactive caspases at specific sites.
- Degradation of essential cellular proteins by the executing caspases with their protease activity.
  - *Death receptors*: Fas/CD95, DR4/DR5, DR3, and TNFR (Tumor Necrosis Factor Receptor).
  - *Adaptors*: FADD (Fas-associated death domain protein) and TRADD (TNFR-associated death domain protein).
  - *Activation*: Binding of death ligands (FasL/CD95L, TRAIL/APO-2L, APO-3L and TNF) induces trimerization of their receptors, which then recruit adaptors and activate caspases.

Note: TRADD is involved only in the coupling between

caspases and DR3 or TNFR. This adaptor can also recruit other proteins (web link) to inhibit apoptosis through the NF-kB pathway.

As shown in the above figure, a variety of death ligands (FasL/CD95L, TRAIL, APO-3L and TNF) can induce apoptosis. It is natural to see if they can kill cancer cells without affecting normal cells. TNF was first investigated in the 1980s for cancer therapy, but with disappointing results.

Then CD95L (FasL) was tested in the 1990s. The results were still not satisfactory. Recently, TRAIL has been demonstrated to be highly selective for transformed cells, with minimal effects on normal cells. It could be an effective drug for both cancer and AIDS.

Nucleases splits nucleic acids (DNA and RNA). Based on the substrate type, they are divided into RNase and DNase. RNase catalyzes the hydrolysis of RNA and DNase acts on DNA. They may also be divided into exonuclease and endonuclease. The exonuclease progressively splits off single nucleotides from one end of DNA or RNA. The endonuclease splits DNA or RNA at internal sites.

Phosphatase catalyzes dephosphorylation (removal of phosphate groups).

Example: calcineurin.The immuno suppressive drugs FK506 and Cyclosporin A are the inhibitors of calcineurin.

NFAT, Calcineurin and Immunosuppression

Organ transplantation often elicits immune responses to reject the foreign object. The most widely used drugs to deal with such graft reaction are FK506 and cyclosporin A (CsA). Their direct target is a phosphatase known as calcineurin. CsA combines with a cellular protein called cyclophilin to inhibit calcineurin; FK506 combines with FKBP (FK506 binding protein) to inhibit the phosphatase.

How could inhibition of a phosphatase suppress the immune response? The answer lies in the transport of NFAT (nuclear factor of activated T cells) which is a transcription factor regulating IL-2. NFAT contains both nuclear localization sequence (NLS) and nuclear export sequence (NES), but one of them is buried in the protein interior, inaccessible to importin and exportin. Whether NLS or NES is masked depends on the phosphorylation state of specific serine residues in the regulatory domain. Phosphorylation of these serine residues exposes NES whereas dephosphorylation exposes NLS. In resting cells, the NLS of the cytoplasmic NFAT is masked due to phosphorylation on these serine residues. In stimulated cells, an increase of intracellular calcium ions activates calcineurin, which then dephosphorylates the masking residues. Consequently, NLS is exposed and NFAT can be carried into the nucleus by the importin a/b. Inside the nucleus, NFAT may be re-phosphorylated by protein kinase, exposing its NES and be exported by the exportin Crm1.

- The structure of the FK506-FKBP-calcineurin complex. PDB ID = 1TCO.
- In the normal process, calcineurin may dephosphorylate NFAT, exposing its NLS so that NFAT can move into the nucleus to stimulate the expression of IL-2, inducing a cascade of immune responses. If calcineurin is inhibited, the immune response will be suppressed.
- Lyases catalyze the cleavage of C-C, C-O, C-S and C-N bonds by means other than hydrolysis or oxidation. Common names include decarboxylase and aldolase.
- Isomerases catalyze atomic rearrangements within a molecule. Examples include rotamase, protein disulfide isomerase (PDI), epimerase and racemase.
- Ligases catalyze the reaction which joins two molecules. Examples include peptide synthase, aminoacyl-tRNA synthetase, DNA ligase and RNA ligase.

The IUBMB committee also defines subclasses and sub-subclasses. Each enzyme is assigned an EC (Enzyme Commission) number. For example, the EC number of catalase is EC1.11.1.6. The first digit indicates that the enzyme belongs to oxidoreductase (class 1). Subsequent digits represent subclasses and sub-subclasses.

CATALYTIC MECHANISMS OF ENZYMES

From the energetic point of view, the reason why an enzyme can accelerate a reaction is because it can lower the energy barrier (activation energy) separating the substrate and the reaction product.

For example, the covalent bond energy between two atoms ranges from 50 to 200 kcal/mol, which is far greater than the thermal energy (0.6 kcal/mol) at room temperature. Therefore, the covalent bond is unlikely to break in the absence of external interactions. Enzymes can provide a proper environment to lower the energy barrier.

The exact mechanism of lowering the energy barrier depends on individual systems. RNase A is a very interesting example. This enzyme can cleave an RNA molecule through hydrolysis reaction, but has no effect on DNA.

MEMBRANE PROTEINS

TRANSMEMBRANE PMZROTEINS

G-protein-Coupled Receptors

The major role of G-protein-coupled receptors is to transmit signals into the cell. They are characterized by seven transmembrane segments. This class of membrane proteins can respond to a wide range of agonists, including photon, amines, hormones, neurotransmitters and proteins. Some agonists bind to the extracellular loops of the receptor, others may penetrate into the transmembrane region.

Immunoglobulin Family

This protein family transmit signals for immune responses. Each of them contains one or two transmembrane segments. The immunoglobulin (i.e., antibody) consists of four peptide chains: two heavy chains and two light chains. V_H and C_H represent the variable and conserved domains of the heavy chain, respectively; V_L and C_L denote those for the light chain.

Variable domains are the binding site of antigens. To send signals into the cell, this class of proteins are typically coupled to non-receptor protein tyrosine kinase located on the cytosolic face of the plasma membrane.

Transport Proteins

According to the Transport Commission (TC) system, all transport proteins may be divided into the following classes:

- Channels/Pores
  Examples: ion channels, porins
- Electrochemical Potential-driven transporters
  Examples: uniporters, symporters and antiporters
- Primary Active Transporters
  Examples: Na^+-K^+ ATPase, H^+ ATPase, ABC Transporter Superfamily and CFTR channel.
- Group Translocators
- Transport Electron Carriers
- Auxiliary Transport Proteins
- Incompletely Characterized

CHANNELS/PORES

Ion Channels [I]

Ion channels are a special class of proteins that conduct small ions such as Na^+, K^+, Ca^{2+} or Cl^-. The opening and closing of ion channels may depend on membrane voltage or the binding of molecules (the ligand). Its mechanism remains unclear. However, the mystery of their selectivity has been solved.

Ion Selectivity Potassium Channels

By definition, a potassium channel conducts mainly K^+ ions while excluding Na^+ and Ca^{2+} ions. We note that a Na^+ ion is smaller than a K^+ ion. How could a channel conduct K^+ ions while excluding smaller ions? The answer lies in the ion's hydration energy.

The diameter of the potassium channel's selectivity filter (the narrowest region of its channel pore) should be about the same as the size of a Cs^+ ion (diameter = 3.3Å) because Cs^+ is the largest ion which can barely pass through the potassium channel. In a solution, ions are surrounded by water molecules. The first hydration shell of a K^+ or Na^+ ion contains six water molecules.

The diameters of unhydrated Na^+ and K^+ ions are 1.96Å and 2.66Å, respectively. The effective diameter of a water molecule in a hydrated ion is larger than 2Å. Therefore, to pass through the selectivity filter, the K^+ or Na^+ ion must remove four water molecules from its first hydration shell, leaving only two (one at the front and anther at the back). The smaller Na^+ ion requires greater dehydration energy than the K^+ ion, because its nucleus has shorter distance with surrounding water molecules and thus interacting more strongly.

As a result, the Na^+ ion is harder to pass through the potassium channel than the K^+ ion. The observation that lithium (Li^+) is excluded from the potassium channel can also be explained by the same energetic consideration. Divalent cations such as Ca^{2+} should also be excluded because their dehydration energies are much higher than monovalent cations.

Sodium Channels

In the sodium channel, the Na^+ ion is more permeable than the K^+ ion. This is because the selectivity filter of the sodium channel is slightly larger than that of the potassium channel. It is large enough to accommodate a Na^+ ion attached with three water molecules, but not enough for a K^+ ion attached with three water molecules.

Therefore, to pass through the sodium channel, the Na^+ ion needs to remove only three, but the K^+ ion has to remove four, water molecules from its first hydration shell. The required dehydration energy for the K^+ ion is greater than the Na^+ ion.

Calcium Channels

In calcium channels, the permeability of monovalent cations (Na^+ and K^+) is about three orders of magnitude smaller than the Ca^{2+} permeability. This ion selectivity does not seem to involve hydration, because Ca^{2+} is more heavily hydrated than Na^+, and the unhydrated diameters of Ca^{2+} and Na^+ are almost identical. Then, how could calcium channels select Ca^{2+} over Na^+? Although the permeability of monovalent cations in the calcium channel is quite small at normal ionic concentrations, large monovalent cationic current can be observed in the absence of Ca^{2+} and other divalent cations. This suggests that the calcium channel is basically permeable to both divalent and monovalent cations, but the selectivity arises from competition between ions.

The calcium channel may contain a negatively charged binding site to facilitate ion conduction. The monovalent cations simply cannot compete with Ca^{2+} for this binding site. This idea has been demonstrated experimentally. In the calcium channel, if a negatively charged glutamate residue in the pore-lining region is mutated into a positively charged lysine, the calcium channel becomes more permeable to Na^+ than Ba^{2+}. Conversely, in the sodium channel, mutation of a pore-lining lysine residue into glutamate transforms the channel from a Na^+-selective to a Ca^{2+}-selective channel.

Channel Structures

The detailed 3D structures of most ion channels are not known. However, from the hydrophobicity profile of their amino acid sequences, it is possible to obtain the domain structures of most ion channels.

K Channels

There are many types of potassium channels. Most commonly observed K channels are composed of four subunits, each is homologous to the Shaker protein. The hydrophobicity profile indicates that it contains six hydrophobic segments, designated as S1 - S6. These segments are likely to be the transmembrane domains. Other experimental results suggests that the P-region is lining the channel pore.

Na and Ca Channels

A Na or Ca channel consists of a major pore-forming

subunit and possibly other small auxiliary subunits. The major pore-forming subunit is called the α subunit which can be divided into four similar domains. Each domain is analogous to a Shaker protein with six transmembrane segments and a P-region. Thus, an α subunit is sufficient to form an ion channel.

Synaptic Channels

A large class of ion channels are specifically located at the synapses (the junction between nerve cells).

Porin

The nonpolar interior of membranes limits the free diffusion of polar molecules. In most membranes various proteins serve as channels and pumps which regulate the movement of ions, metabolites and even water across the membrane. Due to the specificity of these proteins there is a high degree of control over what passes through the membrane.

Several membranes such as the outer membranes of mitochondria, chloroplasts and bacteria, however, do not need the same high degree of control over what gets across. Rather than synthesizing many selective proteins to carry polar molecules across, it appears that cells make only a few types of transporters, like the porin pictured below, which let almost anything below a certain size across.

It may simply be more cost effective to put a few porins in these membranes then many more selective proteins. These "leaky" membranes are still effective barriers to the movement of large molecules such as proteins.

Below are two views of a model of porin from the bacterium *Rhodopseudomonas blastica.* In the left image the model is shown from the side as if you are in the middle of the membrane.

The height of the protein is about the same as the thickness of the membrane in which it sits. The right image shows the model from the end where the central "hole" through which ions and metabolites pass is clearly visible.

Regions of secondary structure are shown in yellow (beta sheet), pink (alpha helix) and blue (turn).

The strands of colour follow the backbone of the molecule. If all of the atoms and bonds in the protein were visualized the image would be much too complicated.Why is the tertiary structure of porin described as a beta barrel?

Electrochemical Potential-driven Transporters

Uniporter-Catalyzed Transport

We begin our discussion of membrane transport proteins with the simplest type, which catalyze uniport transport. The plasma membrane of most cells contains several uniporters that enable amino acids, nucleosides, sugars, and other small molecules to enter and leave cells down their concentration gradients.

Similar to enzymes, uniporters accelerate a reaction that is already thermodynamically favored, and the movement of a substance across a membrane down its concentration gradient will have the same negative ΔG value whether or not a protein transporter is involved. This type of movement sometimes is referred to as facilitated transport (or facilitated diffusion).

Many chemical reactions that are thermodynamically favored will not occur unless an appropriate enzyme is present; such is also the case with movement of hydrophilic molecules across biological membranes. Unlike the substrates of enzymatic reactions, however, transported substances undergo no chemical change during movement across a membrane.

Three properties of uniporter-catalyzed movement of glucose and other small hydrophilic molecules across a membrane distinguish this type of transport from passive diffusion:

- The rate of facilitated transport by uniporters is far higher than predicted by Fick's equation describing passive diffusion. Because the transported molecules never enter the hydrophobic core of the phospholipid bilayer, the partition coefficient K is irrelevant.

- Transport is specific. Each uniporter transports only a single species of molecule or a single group of closely related molecules.
- Transport occurs via a limited number of uniporter molecules, rather *than throughout the phospholipid bilayer.* Consequently, there is a maximum transport rate V_{max} that is achieved when the concentration gradient across the membrane is very large and each uniporter is working at its maximal rate.

Cotransport by Symporters and Antiporters

Besides ATP-powered pumps, cells have a second, discrete class of proteins that import or export ions and small molecules, such as glucose and amino acids, against a concentration gradient. These proteins use the energy stored in the electrochemical gradient of Na^+ or H^+ ions to power the uphill movement of another substance, a process called cotransport.

For instance, the energetically favored movement of a Na^+ ion (the "cotransported" ion) into a cell across the plasma membrane, driven both by its concentration gradient and by the transmembrane voltage gradient can be coupled obligatorily to movement of the "transported" molecule (e.g., glucose) against its concentration gradient. When the transported molecule and cotransported ion move in the same direction, the process is called symport; when they move in opposite directions, the process is called antiport.

Primary Active Transporters

Active Transport by ATP-Powered Pumps

We turn now to the ATP-powered pumps that transport ions and various small molecules against their concentration gradients. The general structures of the four principal classes of these transport proteins are depicted in figure and their properties are summarized in table. Note that the P, F, and V classes transport ions only, whereas the ABC superfamily class transports small molecules as well as ions.

P-class ion pumps containa transmembrane catalytic á subunit, which contains an ATP-binding site, and usually a smaller β subunit, which may have regulatory functions. Many of these pumps are tetramers composed of twoα and two β subunits. During the transport process, at least one of the á subunits is phosphorylated (hence the label "P"), and the transported ions are thought to move through the phosphorylated subunit.

This class includes the Na^+/K^+ ATPase in the plasma membrane, which maintains the Na^+ and K^+ gradients typical of animal cells, and several Ca^{2+} ATPases, which pump Ca^{2+} ions out of the cytosol into the external medium or into the lumen of the sarcoplasmic reticulum (SR) of muscle cells.

Another member of the P class, found in acid-secreting cells of the mammalian stomach, transports protons (H^+ ions) out of and K^+ ions into the cell. The H^+ pump that maintains the membrane electric potential in plant, fungal, and bacterial cells also belongs to this class.

The structures of *F-class* and *V-class ion pumps* are similar to each other but unrelated to and more complicated than P-class pumps. F- and V-class pumps contain at least three kinds of transmembrane proteins and five kinds of extrinsic polypeptides that form the cytosolic domain.

Several of the transmembrane and extrinsic subunits in F-class and V-class pumps exhibit sequence homology, and each pair of homologous subunits is thought to have evolved from a common polypeptide. All known V and F pumps transport only protons in a process that does not involve a phosphoprotein intermediate. V-class pumps generally function to maintain the low pH of plant vacuoles and of lysosomes and other acidic vesicles in animal cells by using the energy released by ATP hydrolysis to pump protons from the cytosolic to the exoplasmic face of the membrane against the proton electrochemical gradient. F-class pumps are found in bacterial plasma membranes and in mitochondria and chloroplasts. In contrast to V pumps, they generally function to power the synthesis of ATP from ADP and P_i by movement of protons from the exoplasmic to the cytosolic face of the membrane down the proton electrochemical gradient. The final class of ATP-powered transport proteins is larger and more diverse than the other classes. Referred to as the *ABC (ATP-binding cassette) superfamily,* this class includes more than 100 different transport proteins found in organisms ranging from bacteria to humans.

Each ABC protein is specific for a single substrate or group of related substrates including ions, sugars, peptides, polysaccharides, and even proteins. All ABC transport proteins share a common organization consisting of four "core" domains: two transmembrane (T) domains, forming the passageway through which transported molecules cross the membrane, and two cytosolic ATP-binding (A) domains. In some ABC proteins, the core domains are present in four separate polypeptides; in others, the core domains are fused into one or two multidomain polypeptides.

All classes of ATP-powered pumps have one or more binding sites for ATP, and these are always on the cytosolic face of the membrane. Although these proteins are often called ATPases, they normally do not hydrolyze ATP into ADP and P_i unless ions or other molecules are simultaneously transported.

Because of the tight coupling between ATP hydrolysis and transport, the energy stored in the phosphoanhydride bond is not dissipated. Thus ATP-powered transport proteins are able to collect the free energy released during ATP hydrolysis and use it to move ions or other molecules uphill against a potential or concentration gradient.

The energy expended by cells to maintain the concentration gradients of Na^+, K^+, H^+, and Ca^{2+} across the plasma and intracellular membranes is considerable. In nerve and kidney cells, for example, up to 25 per cent of the ATP produced by the cell is used for ion transport; in human erythrocytes, up to 50 per cent of the available ATP is used for this purpose.

In cells treated with poisons that inhibit the aerobic production of ATP (e.g., 2,4-dinitrophenol), the ion concentration inside the cell gradually approaches that of the exterior environment as the ions move through plasma membrane channels down their electric and concentration gradients.

Eventually treated cells die: partly because protein synthesis requires a high concentration of K^+ ions and partly because in the absence of a Na^+ gradient across the cell membrane, a cell cannot import certain nutrients such as amino acids. Studies on the effects of such poisons provided early evidence for the existence of ion pumps. In this section, we discuss in some detail examples of the P, V, and ABC classes of ATP-powered pumps.

Plasma-Membrane Ca^{2+} ATPase Exports Ca^{2+} Ions from Cells.

In order for Ca^{2+} to function in intracellular signaling, its cytosolic concentration usually must be kept below 0.1 – 0.2 μM. (Although some cytosolic Ca^{2+} is bound to negatively charged groups, it is the concentration of *free,* unbound Ca^{2+} that is critical to its signaling function.) The plasma membranes of animal, yeast, and probably plant cells contain Ca^{2+} ATPases that transport Ca^{2+} out of the cell against its electrochemical gradient.

These P-class ion pumps help maintain the concentration of free Ca^{2+} ions in the cytosol at a low level. In addition to a catalytica subunit containing an ATP-binding site, as found in other P-class pumps, plasmamembrane Ca^{2+} ATPases also contain the Ca^{2+}-binding regulatory protein calmodulin.

A rise in cytosolic Ca^{2+} induces the binding of Ca^{2+} ions to calmodulin, which triggers an allosteric activation of the Ca^{2+} ATPase; as a result, the export of Ca^{2+} ions from the cell accelerates, and the original low cytosolic concentration of free Ca^{2+} is restored rapidly.

Muscle Ca^{2+} ATPase Pumps Ca^{2+} Ions from the Cytosol into the Sarcoplasmic Reticulum.

Besides the plasma-membrane Ca^{2+} ATPase, muscle cells contain a second, different Ca^{2+} ATPase that transports Ca^{2+} from the cytosol into the lumen of the sarcoplasmic reticulum

(SR), an internal organelle that concentrates and stores Ca^{2+} ions. As discussed the SR and its calcium pump (referred to as the *muscle calcium pump*) are critical in muscle contraction and relaxation: release of Ca^{2+} ions from the SR into the muscle cytosol causes contraction, and the rapid removal of Ca^{2+} ions from the cytosol by the muscle calcium pump induces relaxation.

Because the muscle calcium pump constitutes more than 80 per cent of the integral protein in SR membranes, it is easily purified and characterized. Each transmembrane catalytica subunit has a molecular weight of 100,000 and transports two Ca^{2+} ions per ATP hydrolyzed. In the cytosol of muscle cells, the free Ca^{2+} concentration ranges from 10^{-7} M (resting cells) to more than 10^{-6} M (contracting cells), whereas the *total* Ca^{2+} concentration in the SR lumen can be as high as 10^{-2} M.

Sites on the cytosolic surface of the muscle calcium pump have a very high affinity for Ca^{2+} ($K_m = 10^{-7}$ M), allowing the pump to transport Ca^{2+} efficiently from the cytosol into the SR against the steep concentration gradient.

The concentration of free Ca^{2+} within the sarcoplasmic reticulum is actually much less than the total concentration of 10^{-2} M. Two soluble proteins in the lumen of SR vesicles bind Ca^{2+} and serve as a reservoir for intracellular Ca^{2+}, thereby reducing the concentration of free Ca^{2+} ions in the SR vesicles, and consequently decreasing the energy needed to pump Ca^{2+} ions into them from the cytosol.

The activity of the muscle Ca^{2+} ATPase is so regulated that if the free Ca^{2+} concentration in the cytosol becomes too high, the rate of calcium pumping increases until the cytosolic Ca^{2+} concentration is reduced to less than 1 μM. Thus in muscle cells, the calcium pump in the SR membrane can supplement the activity of the plasma-membrane pump, assuring that the cytosolic concentration of free Ca^{2+} remains below 1 μM.

The current model of the mechanism of action of the Ca^{2+} ATPase in the SR membrane is outlined in Figure. Coupling of ATP hydrolysis with ion pumping involves several steps that must occur in a defined order.When the protein is in one conformation, termed *E1*, two Ca^{2+} ions bind in sequence to high-affinity sites on the cytosolic surface (step 1). Then an ATP binds to its site on the cytosolic surface; in a reaction requiring that a Mg^{2+} ion be tightly complexed to the ATP, the bound ATP is hydrolyzed to ADP and the liberated phosphate is transferred to a specific aspartate residue in the protein, forming a high-energy acyl phosphate bond, denoted by E1~P (step 2).

The protein then changes its conformation to E2 – P, generating two lowaffinity Ca^{2+}-binding sites on the exoplasmic surface, which faces the SR lumen; this conformational change simultaneously propels the two Ca^{2+} ions through the protein to these sites (step 3) and inactivates the high-affinity Ca^{2+}-binding sites on the cytosolic face.

The Ca^{2+} ions then dissociate from the exoplasmic surface of the protein (step 4). Following this, the aspartyl-phosphate bond in E2 – P is hydrolyzed, causing E2 to revert to E1, a change that inactivates the exoplasmic-facing Ca^{2+}-binding sites and regenerates the cytosolicfacing Ca^{2+}-binding sites (step 5).

Thus phosphorylation of the muscle calcium pump by ATP favors conversion of E1 to E2, and dephosphorylation favors the conversion of E2 to E1. While only E2 – P, not E1~P, is actually hydrolyzed, the free energy of hydrolysis of the aspartyl-phosphate bond in E1~P is greater than that for E2 – P. The reduction in free energy of the aspartyl-phosphate bond in E2 – P, relative to E1~P, can be said to power the E1 E2 conformational change. The affinity of Ca^{2+} for the cytosolic-facing binding sites in E1 is a thousandfold greater than the affinity of Ca^{2+} for the exoplasmic-facing sites in E2; this difference enables the protein to transport Ca^{2+} unidirectionally from the cytosol, where it binds tightly to the pump, to the exoplasm, where it is released.

For instance, the muscle calcium pump has been isolated with phosphate linked to an aspartate residue, and spectroscopic studies have detected slight alterations in protein conformation during the E1 E2 conversion.

The membrane-spanningα helices are thought to form the passageway through which Ca^{2+} ions move. The bulk of the subunit consists of cytosolic globular domains that are involved in ATP binding, phosphorylation of aspartate, and energy transduction. These domains are connected by "stalks" to the membrane-embedded domain.

As noted previously, all P-class ion pumps, regardless of which ion they transport, are phosphorylated during the transport process. The amino acid sequences around the phosphorylated aspartate in the catalytic subunit are highly conserved in all proteins of this type.

Thus the mechanistic model probably is generally applicable to all these ATP-powered ion pumps. In addition, the α subunits of all the P pumps examined to date have a similar molecular weight and, as deduced from their amino acid sequences derived from cDNA clones, have a similar arrangement of transmembrane α helices. These findings strongly suggest that all these proteins evolved from a common precursor, although they now transport different ions.

Na^+/K^+ ATPase Maintains the Intracellular Na^+ and K^+ Concentrations in Animal Cells.

A second P-class ion pump that has been studied in considerable detail is the Na^+/K^+ ATPase present in the plasma membrane of all animal cells. This ion pump is a tetramer of subunit composition $\alpha_2\beta_2$. (Classic Experiment 15.1 describes the discovery of this enzyme.) The β polypeptide is required for newly synthesized α subunits to fold properly in the endoplasmic reticulum but apparently is not involved directly in ion pumping.

The α subunit is a 120,000-MW nonglycosylated polypeptide whose amino acid sequence and predicted membrane structure are very similar to those of the muscle Ca^{2+} ATPase. In particular, the Na^+/K^+ ATPase has a stalk on the cytosolic face that links domains containing the ATP-binding site and the phosphorylated aspartate to the membrane-embedded domain. The overall process of transport moves three Na^+ ions out of and two K^+ ions into the cell per ATP molecule split.

Several lines of evidence indicate that the Na^+/K^+ ATPase is responsible for the coupled movement of K^+ and Na^+ into and out of the cell, respectively. For example, the drug *ouabain*, which binds to a specific region on the exoplasmic surface of the protein and specifically inhibits its ATPase activity, also prevents cells from maintaining their Na^+/K^+ balance.

Any doubt that the Na^+/K^+ ATPase is responsible for ion movement was dispelled by the demonstration that the enzyme, when purified from the membrane and inserted into liposomes, propels K^+ and Na^+ transport in the presence of ATP.

The mechanism of action of the Na^+/K^+ ATPase, outlined in figure is similar to that of the muscle calcium pump, except that ions are pumped in both directions across the membrane. In its E1 conformation, the Na^+/K^+ ATPase has three high-affinity Na^+-binding sites and two low-affinity K^+-binding sites on the cytosolic-facing surface of the protein.

The K_m for binding of Na^+ to these cytosolic sites is 0.6 mM, a value considerably lower than the intracellular Na^+ concentration of H"12 mM; as a result, Na^+ ions normally will fill these sites. Conversely, the affinity of the cytosolic K^+-binding sites is low enough that K^+ ions, transported inward through the protein, dissociate from E1 into the cytosol despite the high intracellular K^+ concentration. During the E1 E2 transition, the three bound Na^+ ions move outward through the protein.

Transition to the E2 conformation also generates two high-affinity K^+ sites and three low-affinity Na^+ sites on the exoplasmic face. Because the K_m for K^+ binding to these sites (0.2 mM) is considerably lower than the extracellular K^+ concentration (4 mM), these sites will fill quickly with K^+ ions. In contrast, the three Na^+ ions, transported outward through the protein, will dissociate into the extracellular medium from the low-affinity Na^+ sites on the exoplasmic surface despite the high extracellular Na^+ concentration. Similarly, during the E2 E1 transition, the two bound K^+ ions are transported inward.

V-Class H^+ ATPases Pump Protons across Lysosomal and Vacuolar MembranesAll V-class ATPases transport H^+ ions only. These proton pumps, present in the membranes of lysosomes, endosomes, and plant vacuoles, function to acidify the lumen of these organelles. The acidity of the lysosomal lumen, usually H"4.5 – 5.0, can be measured precisely in living cells by use of particles labeled with a pH-sensitive fluorescent dye. Cells phagocytose these particles and transfer them to the lysosomes.

The ability of different wavelengths of visible light to excite fluorescence is highly dependent on pH, and the lysosomal pH can be calculated from the spectrum of the fluorescence emitted. Maintenance of the 100-fold or more proton gradient between the lysosomal lumen (pH H"4.5 – 5.0) and the cytosol (pH H"7.0) depends on ATP production by the cell.

The ATP-powered proton pumps in lysosomal and vacuolar membranes have been isolated, purified, and incorporated into liposomes.

As illustrated in figure these V-class proton pumps contain two discrete domains: a cytosolic-facing hydrophilic domain (V_1) composed of five different polypeptides and a transmembrane domain (V_0) containing 9 – 12 copies of proteolipid c, one copy of protein b, and one copy of protein a. The subunit composition of the cytosolic domain is $\alpha_3\beta_3\gamma\delta\varepsilon$; the α and β subunits contain the sites where ATP binding and hydrolysis occur.

Each transmembrane c subunit is thought to span the membrane two times; the c and a subunits together form the proton-conducting channel. Unlike P-class ion pumps, the V-class H^+ ATPases are not phosphorylated and dephosphorylated during proton transport. Similar V-class ATPases are found in the plasma membrane of certain acid-secreting cells. These include osteoclasts, bone-resorbing macrophagelike cells, which bind to a bone and seal off a small segment of extracellular space between the plasma membrane and the surface of the bone. HCl secreted into this space by osteoclasts dissolves the calcium phosphate crystals that give bone its rigidity and strength.

Another example is the mitochondria-rich epithelial cells

lining the toad bladder; the apical plasma membrane of these cells contain many V-class H^+ ATPases, which function to acidify the urine. As we discuss later, the membrane of plant vacuoles contains two proton pumps: a typical V-class H^+ ATPase and another one that utilizes the energy released by hydrolysis of inorganic pyrophosphate (PP_i) to pump protons into the vacuole. This PP_i-hydrolyzing proton pump, believed to be unique to plants, has an amino acid sequence different from any other ion-transporting proteins

ATP-powered proton pumps cannot acidify the lumen of an organelle (or the extracellular space) by themselves. The reason for this is that pumping of protons would rapidly cause a buildup of positive charge on the exoplasmic face of the membrane on the inside of the vesicle membrane and a corresponding buildup of negative charges on the cytosoiic face. In other words, the pump would generate a voltage across the membrane, exoplasmic face positive, which would prevent movement of protons into the vesicle before a significant H^+ concentration gradient had been established. In fact, this is the way that H^+ pumps generate an insidenegative potential across plant and yeast plasma membranes. In order for an organelle lumen or an extracellular space (e.g., the outside of an osteoclast) to become acidic, movement of H^+ up its concentration gradient must be accompanied by:

- Movement of an equal number of anions in the same direction or
- Movement of equal numbers of a different cation in the opposite direction. The first process occurs in lysosomes and plant vacuoles whose membranes contain V-class H^+ ATPases and ion channels through which accompanying anions (e.g., Cl^-) move. The second occurs in the lining of the stomach, which contains a P-class H^+/K^+ ATPase that pumps one H^+ outward and one K^+ inward.

The ABC Superfamily Transports a Wide Variety of Substrates

As noted earlier, all members of the very large and diverse

ABC superfamily of transport proteins contain two transmembrane (T) domains and two cytosolic ATP-binding (A) domains. The T domains, each built of six membrane-spanning á helices, form the pathway through which the transported substance (substrate) crosses the membrane and determine the substrate specificity of each ABC protein.

The sequence of the A domains is H"30 to 40 per cent homologous in all members of this superfamily, indicating a common evolutionary origin. Some ABC proteins also contain a substrate-binding subunit or regulatory subunit.

Bacterial Plasma-Membrane Permeases

The plasma membrane of many bacteria contain numerous *permeases* that belong to the ABC superfamily. These proteins use the energy released by hydrolysis of ATP to transport specific amino acids, sugars, vitamins, or even peptides into the cell.

Since bacteria frequently grow in soil or pond water where the concentration of nutrients is low, these ABC transport proteins allow the cells to concentrate amino acids and other nutrients in the cell against a substantial concentration gradient. Bacterial permeases generally are *inducible;* that is, the quantity of a transport protein in the cell membrane is regulated by both the concentration of the nutrient in the medium and the metabolic needs of the cell.

In *E. coli* histidine permease, a typical bacterial ABC protein, the two transmembrane domains and two cytosolic ATP-binding domains are formed by four separate subunits. In gram-negative bacteria such as *E. coli,* which have an outer membrane, a soluble histidine-binding protein in the periplasmic space assists in transport. This soluble protein binds histidine tightly and directs it to the T subunits, through which histidine crosses the membrane powered by ATP hydrolysis.

Mutant *E. coli* cells that are defective in any of the histidine-permease subunits or the soluble binding protein are unable to transport histidine into the cell, but are able to transport other amino acids whose uptake is facilitated by other transport proteins. Such genetic analyses provide strong evidence that histidine permease and similar ABC proteins function to transport solutes into the cell.

Mammalian MDR Transport Proteins

A series of rather unexpected observations led to discovery of the first eukaryotic ABC protein. Oncologists noted tnat tumor cells often became simultaneously resistant to several chemotherapeutic drugs with unrelated chemical structures; similarly, cell biologists observed that cultured cells selected for resistance to one toxic substance (e.g., colchicine, a microtubule inhibitor) frequently became resistant to several other drugs, including the anticancer drug adriamycin. Subsequent studies showed that this resistance is due to enhanced expression of a*multidrug-resistance (MDR) transport protein* known as *MDR1*.

In this member of the ABC superfamily, all four domains are "fused" into a single 170,000-MW protein. This protein uses the energy derived from ATP hydrolysis to*export* a large variety of drugs from the cytosol to the extracellular medium. The *Mdr1* gene is frequently amplified in multidrug-resistant cells, resulting in a large overproduction of the MDR1 protein.

Most drugs transported by MDR1 are small hydrophobic molecules, which diffuse from the culture medium across the plasma membrane into the cell. The ATP-powered export of such drugs from the cytosol by MDR1 means a much higher extracellular drug concentration is required to kill cells.

That MDR1 is an ATP-powered small-molecule pump has been demonstrated with liposomes containing the purified protein. The ATPase activity of these liposomes is enhanced by different drugs in a dose-dependent manner corresponding to their ability to be transported by MDR1.

Not only does MDR1 transport a varied group of molecules, but all these substrates compete with one another for transport by MDR1. Although the mechanism of action of MDR1-assisted transport has not been definitively demonstrated, the *flippase model*, is a likely candidate. Substrates of MDR1 are primarily planar, lipid-soluble molecules with one or more positive charges, and they move spontaneously from the cytosol into the cytosolic-facing leaflet of the plasma membrane.

The hydrophobic portion of a substrate molecule is oriented toward the hydrophobic core of the membrane, and the charged portion toward the polar cytosolic face of the membrane and is still in the cytosol. The substrate diffuses laterally until encountering and binding to a site on the MDR1 protein that is within the bilayer. The protein then "flips" the charged substrate molecule into the exoplasmic leaflet, an energetically unfavorable reaction powered by the coupled ATPase activity of MDR1.

Once in the exoplasmic face, the substrate diffuses into the aqueous phase on the outside of the cell. Support for the flippase model of transport by MDR1 comes from MDR2, a homologous protein present in the region of the liver cell plasma membrane that faces the bile duct.

MDR2 has been shown to flip phospholipids from the cytosolic-facing leaflet of the plasma membrane to the exoplasmic leaflet, thereby generating an excess of phospholipids in the exoplasmic leaflet; these phospholipids peel off into the bile duct and form an essential part of the bile.

An alternative *pump model* also has been proposed for MDR1. According to this model, drug molecules in the cytosol bind directly to a single small-molecule binding site on the cytosolic face of the MDR1 protein; subsequent ATP hydrolysis powers movement of the bound drug through the protein to the aqueous phase on the outside of the cell by a mechanism similar to that of other ATP-powered pumps.

MDR1 protein is expressed in abundance in the liver, intestines, and kidney — sites from which natural toxic products are removed from the body. Thus the natural function of MDR1 may be to transport a variety of natural and metabolic toxins into the bile, intestinal lumen, or forming urine.

During the course of its evolution, MDR1 appears to have coincidentally acquired the ability to transport drugs whose structures are similar to those of these toxins. Tumors derived from these cell types, such as hepatomas (liver cancers), frequently are resistant to virtually all chemotherapeutic agents and thus difficult to treat, presumably because the tumors exhibit increased expression of the MDR1 or MDR2 proteins.

Cystic Fibrosis Transmembrane Regulator (CFTR) Protein

Discovery of another ABC transport protein came from studies of cystic fibrosis (CF), the most common lethal autosomal recessive genetic disease of Caucasians. This disease is caused by a mutation in the *CFTR* gene, which encodes a chloride-channel protein that is regulated by cyclic AMP (cAMP), an intracellular second messenger.

These Cl^{-} channels are present in the apical plasma membranes of epithelial cells in the lung, sweat glands, pancreas, and other tissues. An increase in cAMP stimulates Cl^{-} transport by such cells from normal individuals, but not from CF individuals who have a defective CFTR protein.

The sequence and predicted structure of the encoded CFTR protein, based on analysis of the cloned gene, are very similar to those of MDR1 protein except for the presence of an additional domain, the regulatory (R) domain, on the cytosolic face. The Cl^{-}-channel activity of CFTR protein clearly is enhanced by binding of ATP.

Moreover, cAMP activates a protein kinase that phosphorylates, and thereby activates, CFTR. When purified CFTR protein is incorporated into liposomes, it forms Cl^{-} channels with properties similar to those in normal epithelial cells.

And when the wild-type CFTR protein is expressed by recombinant techniques in cultured epithelial cells from CF patients, the cells recover normal Cl^{-}-channel activity. This latter result raises the possibility that gene therapy might reverse the course of cystic fibrosis.

Since CFTR protein is similar to MDR1 in structure, it may also function as an ATP-powered pump of some as-yet unidentified molecule. In any case, much remains to be learned about this fascinating class of ABC transport proteins.

MEMBRANE-ATTACHED PROTEINS

Membrane-attached proteins can be divided into three groups:

- Bound to the cytosolic face of the plasma membrane by myristate;
- Bound to the cytosolic face of the plasma membrane by farnesyl;
- Bound to the extracelluar face of the plasma membrane by glycosylphosphatidylinositol.

MISCELLANEOUS PROTEINS

In addition to enzymes and membrane proteins, there are many other proteins with various cellular functions. This page is not intended to be comprehensive, but lists only a few important examples.

CHAPERONES

Involved in protein folding.

The Small Heat Shock Proteins and Their Role in Human Disease

Small heat shock proteins (sHSPs) function as molecular chaperones, preventing stress induced aggregation of partially denatured proteins and promoting their return to native conformations when favorable conditions pertain.

Sequence similarity between sHSPs resides predominately in an internal stretch of residues termed the crystallin domain, a region usually flanked by two extensions. The poorly conserved N-terminal extension influences oligomer construction and chaperone activity, whereas the flexible C-terminal extension stabilizes quaternary structure and enhances protein/substrate complex solubility. sHSP polypeptides assemble into dynamic oligomers which undergo subunit exchange and they bind a wide range of cellular substrates.

As molecular chaperones, the sHSPs protect protein

structure and activity, thereby preventing disease, but they may contribute to cell malfunction when perturbed. For example, sHSPs prevent cataract in the mammalian lens and guard against ischemic and reperfusion injury due to heart attack and stroke.

On the other hand, mutated sHSPs are implicated in diseases such as desmin-related myopathy and they have an uncertain relationship to neurological disorders including Parkinson's and Alzheimer's disease. This review explores the involvement of sHSPs in disease and their potential for therapeutic intervention.

Within the molecular chaperone family, sHSPs constitute a structurally divergent group characterized by a conserved sequence of 80–100 amino acid residues termed the -crystallin domain. The -crystallin domain, duplicated in the unusual example of parasitic flatworms (Platyhelminthes), is located toward a highly flexible, variable, C-terminal extension, and is usually preceded by a poorly conserved N-terminal region. The molecular mass of sHSP subunits ranges from 12 to 43 kDa, and they assemble into large, dynamic complexes up to 1 MDa. sHSP secondary structure is dominated by:

- -strands with limited
- -helical content,
- -sheets within the crystallin domain mediate dimer formation.

Crystallization of two sHSPs has contributed significantly to the description of oligomerization, quaternary structure, subunit exchange, and chaperone activity. Characterization of a highly conserved arginine is also an important outcome of crystallization and related studies because mutation of this residue has profound effects on sHSP function and contributes to certain diseases.

The sHSPs are molecular chaperones, storing aggregation prone proteins as folding competent intermediates and conferring enhanced stress resistance on cells by suppressing aggregation of denaturing proteins, actions associated with oligomerization and subunit exchange.

Functional studies of the sHSPs are more limited than for

other chaperones, but this is changing as the application of genomics and proteomics reveals sHSP characteristics and their medical importance emerges. In this context, 10 sHSPs, termed HspB1–10, many of which are constitutively present at high levels in muscle and implicated in disease, are found in humans.

Intracellular quantities and cellular localizations of sHSPs change in response to development, physiological stressors such as anoxia/hypoxia, heat and oxidation, and in relation to pathological status. sHSPs interact with many essential cell structures and it follows from such promiscuity that functional disruption and inappropriate association of these molecular chaperones with substrates will foster disease.

Therefore, this review considers the role of sHSPs in several human medical conditions and it ends with a discussion of their therapeutic potential.

sHSPs were described previously as the 'forgotten chaperones', but this is no longer true. Two sHSPs have been crystallized, opening the door to more informed interpretation of results obtained by site-directed mutagenesis and other molecular probing. The functions of sHSP domains and individual amino acid residues are becoming clearer, as is the molecular basis of oligomerization.

The implications of oligomer assembly and disassembly as chaperoning prerequisites are under study, sHSP substrates have been identified, and the role of ATP-dependent chaperones in substrate release and refolding revealed. sHSPs operate in the front lines of cell defence, protecting proteins during stress and providing opportunities for salvage.

As molecular chaperones, sHSPs have the potential to guard cells from disease, but when perturbed or as residents of aberrant cells, they may promote disease. For example, sHSPs defend against ischemia, oxidative damage and apoptosis, but post-translational modifications and gene mutations cause cataract and desmin-related myopathies.

Disease involvement suggests therapeutic exploitation of sHSPs, but this remains poorly explored, as is generally true for other HSPs.

However, as sHSPs are better understood, opportunities for disease prevention and treatment become more apparent, and this, along with their fundamental importance in stress physiology, means that sHSPs will not be forgotten for some time to come.

The Role of Structural Disorder in the Function of RNA and Protein Chaperones

Chaperones are highly sophisticated protein machines that assist the folding of RNA molecules or other proteins. Their function is generally thought to require a fine-tuned and highly conserved structure: despite the recent recognition of the widespread occurrence of structural disorder in the proteome, this structural trait has never been generally considered in molecular chaperones.

In this review we give evidence for the prevalence of functional regions without a well-defined 3-D structure in RNA and protein chaperones.

By considering a variety of individual examples, we suggest that the structurally disordered chaperone regions either function as molecular recognition elements that act as solubilizers or locally loosen the structure of the kinetically trapped folding intermediate via transient binding to facilitate its conformational search.

The importance of structural disorder is underlined by a predictor of natural disordered regions, which shows an extremely high proportion of such regions, unparalleled in any other protein class, within RNA chaperones: 54.2per cent of their residues fall into disordered regions and 40per cent fall within disordered regions longer than 30 consecutive residues.

Structural disorder also prevails in protein chaperones, for which frequency values are 36.7per cent and 15per cent, respectively. In keeping with these and other details, a novel "entropy transfer" model is presented to account for the mechanistic role of structural disorder in chaperone function.—Tompa, P., Csermely, P. The role of structural disorder in the function of RNA and protein chaperones.

Intrinsically disordered/unstructured proteins and protein

domains exist and function without a well-defined 3-dimensional fold. Although these proteins defy the classical structure-function paradigm, they are not exceptions that violate the rule: they are rather common in living organisms and fulfill essential cellular functions.

The prevalence of intrinsic disorder has been determined in different functional classes of proteins, and the highest level was found in regulatory, cell signaling, and cancer-associated proteins. Disorder and flexibility have never been comprehensively assessed in chaperones, although sporadic data suggest their presence and functional importance in this protein class.

Chaperones assist RNA molecules or other proteins in reaching their functional 3-dimensional structure. Often, folding of long polypeptide or polynucleotide chains is impeded by kinetic traps (i.e., local energy minima), which are stable enough to halt the folding process for a physiologically significant amount of time. Furthermore, such partially folded or misfolded intermediates tend to undergo irreversible aggregation, which poses a particular danger to the organism.

This folding problem can be overcome by the aid of specific chaperone proteins that prevent aggregation and unfold the intermediate to offer another chance for a folding attempt. Chaperone function is structurally and mechanistically rather demanding, as chaperones have to assist the folding of a wide range of unrelated RNAs/proteins in the crowded intracellular milieu. No wonder the appearance of such highly sophisticated protein machines was a critical early evolutionary invention.

The intriguing aspect of disorder in chaperone function and evolution is highlighted by a recent suggestion that the first proteins to come into being were RNA chaperones. It is generally accepted that the biosphere once was dominated by organisms in which RNA was used for information storage and catalysis.

Because RNA is especially prone to misfolding, simple proteins with chaperone activity must have conferred a significant selective advantage in a world so highly dependent on RNA function. As the first proteins must have been short and unfolded polypeptides, the presence of disordered sequences in protein chaperones and the recent discovery of fully disordered chaperone-like proteins and structural disorder in RNA chaperones underline the possibility of the critical early contribution, and the likely contemporary significance, of structural disorder in chaperone function.

In this chapter, the experimental evidence for the occurrence and functional role of structural disorder in RNA and protein chaperones is reviewed. We show that disordered segments occur in a wide range of chaperones and such regions are often indispensable in functional terms.

The general occurrence of structural disorder is further demonstrated by the newly developed predictor for protein disorder (PONDR) algorithm: the frequency of disorganized structure is higher among RNA chaperones than in any other protein class, and is also very high for protein chaperones. The model of entropy transfer for disorder in chaperone function is suggested to contain all the individual examples and the very high frequency of structural disorder observed.

Chaperone-Mediated Protein Folding

The folding of most newly synthesized proteins in the cell requires the interaction of a variety of protein cofactors known as molecular chaperones. These molecules recognize and bind to nascent polypeptide chains and partially folded intermediates of proteins, preventing their aggregation and misfolding.

There are several families of chaperones; those most involved in protein folding are the 40-kDa heat shock protein (HSP40; DnaJ), 60-kDa heat shock protein (HSP60; GroEL), and 70-kDa heat shock protein (HSP70; DnaK) families. The availability of high-resolution structures has facilitated a more detailed understanding of the complex chaperone machinery and mechanisms, including the ATP-dependent reaction cycles of the GroEL and HSP70 chaperones. For both of these chaperones, the binding of ATP triggers a critical conformational change leading to release of the bound substrate protein.

Whereas the main role of the HSP70/HSP40 chaperone system is to minimize aggregation of newly synthesized proteins, the HSP60 chaperones also facilitate the actual folding process by providing a secluded environment for individual folding molecules and may also promote the unfolding and refolding of misfolded intermediates.

The basic paradigm of molecular chaperones is that they recognize and selectively bind nonnative, but not native, proteins to form relatively stable complexes. In most cases, the complexes are dissociated by the binding and hydrolysis of ATP. In addition, there are "specific" molecular chaperones that typically are involved in the assembly of particular multiprotein complexes.

Molecular chaperones comprise several highly conserved families of unrelated proteins; many chaperones are also heat shock (stress) proteins. The ubiquitous role of molecular chaperones continues to unfold with more discoveries each year. In the context of in vivo protein folding, chaperones prevent irreversible aggregation of nonnative conformations and keep proteins on the productive folding pathway.

In addition, they may maintain newly synthesized proteins in an unfolded conformation suitable for translocation across membranes and bind to nonnative proteins during cellular stress, among other functions. It is likely that most, if not all, cellular proteins will interact with a chaperone at some stage of their lifetime.

The focus of this review is on the functional contribution of chaperones to in vivo protein folding and assembly, especially those chaperones that are promiscuous, in that they show broad specificity for binding nonnative proteins This review is of necessity selective; the main goal is to furnish an up-to-date overview of the role of the major molecular chaperones involved in protein folding. In view of the vast array of literature on molecular chaperones, and the ease of access to literature citations using the Internet, this article should not be viewed as exhaustive.

Why do we need chaperones? After all, a basic tenet of in

vitro protein folding has been the seminal work of Anfinsen which demonstrated that formation of the native protein from the unfolded state is a spontaneous process determined by the global free energy minimum. The results indicated that the native state of small globular proteins is determined by their amino acid sequence.

However, the experimental conditions necessary to successfully fold many proteins, especially larger ones, in vitro, are very constrictive, usually requiring very low protein concentration and long incubation times and are usually unphysiological (e.g., relatively low temperatures). In contrast, most cells operate at ambient or homeothermically set temperatures (e.g., 37°C) where the hydrophobic effect will be stronger and thus protein denaturation and aggregation will be bigger problems, and the time-frame available for successful folding is short.

Thus there is the need for additional factors for the successful folding of many proteins in vivo. When one considers the crowded cellular environment within a cell, it becomes clear that in vitro folding experiments at low protein concentrations are poor models for what happens in the cell, where a newly synthesized protein is in an environment with little or no "free" water, very high concentrations of other proteins and metabolites, and typically membranes, cytoskeletal elements, and other cellular components.

Thus the need for chaperones:

- To prevent aggregation and misfolding during the folding of newly synthesized chains,
- To prevent nonproductive interactions with other cell components,
- To direct the assembly of larger proteins and multiprotein complexes, and
- During exposure to stresses that cause previously folded proteins to unfold, becomes evident. In the few cases where folding has been studied both in vivo and in vitro, it appears that the folding pathways are similar.

Cells have solved the problem of misfolding and

aggregation, to a considerable extent at least, through the participation of molecular chaperones in the in vivo folding process. Many investigations in the past few years have confirmed the critical role of molecular chaperones in protein folding in the cell. Although much has been learned about the function of chaperones in protein folding, and the general outline of the process is thought to be understood, there are still many important unresolved issues, and new chaperones and cochaperones are still being discovered.

The molecular chaperones involved in the folding of newly synthesized proteins recognize nonnative substrate proteins predominantly via their exposed hydrophobic residues. The major chaperone classes are 40-kDa heat shock protein (HSP40; the DnaJ family), 60-kDa heat shock protein [HSP60; including GroEL and the T-complex polypeptide 1 (TCP-1) ring complexes], 70-kDa heat shock protein (HSP70), and 90-kDa heat shock protein (HSP90).

All these chaperones can prevent the aggregation of at least some unfolded proteins. For HSP60 and HSP70, their activity is modulated by the binding and hydrolysis of ATP. The HSP70 (DnaK in *Escherichia coli*) bind to nascent polypeptide chains on ribosomes, preventing their premature folding, misfolding, or aggregation, as well as to newly synthesized proteins in the process of translocation from the cytosol into the mitochondria and the endoplasmic reticulum (ER).

The HSP70 are regulated by HSP40 (DnaJ or its homologs). The HSP60 are large oligomeric ring-shaped proteins known as chaperonins that bind partially folded intermediates, preventing their aggregation, and facilitating their folding and assembly. This family is composed of GroEL-like proteins in eubacteria, mitochondria, and chloroplasts and the TCP-1 (CCT or TRiC) family in the eukaryotic cytosol and the archaea. The HSP60 (GroEL in *E. coli*) are large, usually tetradecameric proteins with a central cavity in which nonnative protein structures bind.

The HSP60 are found in all biological compartments except the ER. The HSP60 are regulated by a cochaperone, chaperonin 10 (cpn10) (GroES in *E. coli*). In addition to preventing aggregation, it has been suggested that HSP60 may permit misfolded structures to unfold and refold. The HSP90 are associated with a number of proteins and play important roles in modulating their activity, most notably the steroid receptors.

A number of other proteins involved in the folding of many newly synthesized proteins are often considered to be molecular chaperones; these include protein disulfide isomerase and peptidyl prolyl isomerase, which catalyze the rearrangement of disulfide bonds and isomerization of peptide bonds around Pro residues, respectively, and are perhaps better considered to be folding catalysts rather than chaperones.

As mentioned previously, there are also a number of more specific chaperones that are involved in the folding/assembly of only one, or a very limited number, of particular substrate proteins.

Chaperones are catalysts in the sense that they transiently interact with their substrate proteins but are not present in the final folded product, and also in that they increase the yield of folded protein. However, there is no good evidence that they actually enhance the spontaneous rate of folding itself, although they may appear to do this by minimizing off-pathway reactions.

A brief perusal of the literature demonstrates that our knowledge of molecular chaperones is growing at an enormous pace. To put these new discoveries in context, a few more general points are worthy of note.

Although in many respects the field of molecular chaperones can now be considered a mature one, in that it has passed its first decade of life, and the broad outlines, at least, are reasonably well established, there are still many outstanding questions. Furthermore, there are many areas of considerable controversy, and many of these relate to fundamental questions.

For example, we do not yet know with certainty whether all newly synthesized proteins interact with chaperones, although it is likely that they do. We certainly do not know much about all the interactions between the various chaperones themselves, as well as with newly synthesized proteins or other chaperone target proteins. There are significant controversies concerning which chaperones interact first with nascent polypeptides, and even whether all nascent polypeptides interact with chaperones. The GroEL family of chaperones has been intensively studied, especially in the context of in vitro protein folding, yet it is not clear just how important a role this family (the cpn60 chaperonins and their TCP-1 eukaryotic homologs) play in the folding of most proteins in the cell.

We are only now beginning to get a picture of the apparently ubiquitous role of the HSP90 family in many critical processes in the cell, especially those involving protein-protein interactions. Recently, several new "accessory" proteins have been discovered, which apparently act as "cochaperones." Again, their significance to protein folding and denaturation in the cell in general is unclear at the present time; they may be highly specialized or may turn out to be critical in a broad range of cellular processes involving chaperones.

Although some of the chaperones clearly are important in preventing protein aggregation, there is as yet no good evidence that chaperones play a role significant in the opposite side of this equation, namely, in solubilizing protein aggregates, although it would seem likely that this may in fact be a function of some chaperones. Even at the level of the specific mechanisms of chaperone function, there are many controversial aspects, and those in the field know there have been some quite rancorous discussions over competing mechanisms. Thus the molecular chaperone field is one in which there are still many outstanding questions, including some quite fundamental ones. Consequently, chaperone scientists are likely to remain busy for a long time to come.

Hsp90: A Specialized but Essential Protein-folding Tool

Cells are equipped with several classes of structurally

unrelated molecular chaperones to ensure both the proper folding of proteins upon synthesis and their refolding under conditions of denaturing stress. Different chaperones follow distinct strategies to achieve the general goal of preventing protein misfolding and aggregation in the highly crowded cellular environment. For example, the monomeric Hsp70 chaperone proteins promiscuously recognize short hydrophobic peptide segments such as those exposed in nascent polypeptide chains.

In contrast, the so-called chaperonins form oligomeric double ring structures capable of enclosing entire proteins or protein domains to promote folding in a sequestered environment. Chaperone proteins often cooperate in folding: a polypeptide that is bound by Hsp70 at an early stage of folding can be transferred to a chaperonin to reach its final native state.

For certain client polypeptides in the eukaryotic cytosol, an alternate pathway has been proposed in which binding to Hsp70 is followed by an interaction with the dimeric chaperone protein Hsp90. Recent advances in understanding the structure and function of Hsp90 now indicate that it uses a protein-folding strategy different from that of either Hsp70 or the chaperonins. Here, we discuss the distinct biochemical mechanisms by which Hsp90 handles its substrate polypeptides and the unique biological functions of this curious protein-folding tool.

Unlike Hsp70, eukaryotic cytosolic Hsp90 does not act generally in nascent protein folding. Hsp90 is distinguished from other chaperones in that most of its known substrates are signal transduction proteins, the classical examples being steroid hormone receptors and signaling kinases.

Because Hsp90 is essential for maintaining the activity of numerous signaling proteins, it plays a key role in cellular signal transduction networks. At a molecular level, Hsp90 binds to substrate proteins, which are in a near native state and thus at a late stage of folding poised for activation by ligand binding or interaction with other factors.

In fulfilling its role, Hsp90 operates as part of a

multichaperone machinery in the cytosol, which includes Hsp70 as mentioned above but also peptidyl-prolyl isomerases and other cochaperones. This complexity of function has made Hsp90 more difficult to investigate than the Hsp70 or chaperonin proteins.

For example, Hsp90 has only been established recently as an ATP-dependent chaperone and the exact functions of the Hsp90 ATPase activity are still being worked out. In the following sections, we will first describe the structure and mechanism of the Hsp90 protein, then focus on its regulation by cochaperones, and finally discuss the biological contribution of Hsp90 to cellular processes including signal transduction, protein degradation, and morphological evolution.

Molecular Chaperones: Biology and Prospects for Pharmacological Intervention

Successful expression of polypeptide gene products requires transcription machinery to generate messenger ribonucleic acids and translation machinery to synthesize polypeptide chains but the protein product is functional, and thus the gene is effectively expressed, only when the polypeptide chain is folded into its native three-dimensional conformation, appropriately localized within or secreted from the cell, and, in many instances, properly assembled into multicomponent complexes.

Although the amino acid sequence alone is sufficient to dictate the native conformation of small proteins in vitro most polypeptides would fail to fold efficiently in the highly concentrated, complex, cellular environment without the assistance of yet another type of machinery. This latter type of machinery involves the so-called molecular chaperones, i.e., proteins adapted to facilitate protein folding. The chaperone concept was first proposed by Ellis and a large, highly active field of research has since developed to investigate the functions and physiological implications of molecular chaperones. Major aspects of this field have been recently reviewed and many details have been compiled in a more comprehensive manner.

There is an inherent problem in the folding of nascent polypeptide chains. Whether synthesized on free ribosomes in the cytoplasm or on ribosomes associated with endoplasmic reticulum (ER) nascent protein chains emerge in a linear manner.

Consequently, hydrophobic stretches and other potentially interactive sites must sometimes wait for downstream sequences to emerge before appropriate folding interactions can be established. Inappropriate interactions can readily lead to poorly reversible conformations and aggregations that reduce the efficiency of native folding reactions. Molecular chaperones reversibly interact with nascent chains to minimize offpathway interactions and increase the yield of native folded protein.

In addition to the folding of nascent polypeptide chains, other cellular processes involve protein folding that requires chaperone participation. One of these is the transport of previously synthesized proteins across cell membranes, as best understood for mitochondrial protein import.

Many mitochondrial proteins are encoded by nuclear genes, synthesized on free ribosomes in the cytoplasm, and subsequently transported across one or both mitochondrial membranes. Proteins are translocated across the mitochondrial membranes in a linear manner as unfolded chains, and there are both cytoplasmic and mitochondrial chaperones that participate in the unfolding and refolding pathways.

Heat Shock Proteins and Cardiovascular Pathophysiology

Heat Shock Proteins and Cardiovascular Pathophysiology. In the eukaryotic cell an intrinsic mechanism is present providing the ability to defend itself against external stressors from various sources.

This defence mechanism probably evolved from the presence of a group of chaperones, playing a crucial role in governing proper protein assembly, folding, and transport. Upregulation of the synthesis of a number of these proteins upon environmental stress establishes a unique defence system to maintain cellular protein homeostasis and to ensure survival of the cell.

In the cardiovascular system this enhanced protein synthesis leads to a transient but powerful increase in tolerance to such endangering situations as ischemia, hypoxia, oxidative injury, and endotoxemia.

These socalled heat shock proteins interfere with several physiological processes within several cell organelles and, for proper functioning, are translocated to different compartments following stress-induced synthesis. In this review we describe the physiological role of heat shock proteins and discuss their protective potential against various stress agents in the cardiovascular system.

Since the first report on the heat-induced appearance of chromosomal puffings in salivary gland tissue of *Drosophila busckii* in 1962, a new research domain has been intensively explored. This research resulted in the discovery of a large number of related proteins and their physiological role in many prokaryotic and eukaryotic organisms, tissues, and individual cells and at the level of subcellular structures.

These proteins were originally called "heat shock proteins" because they were discovered in salivary glands and other tissues of *Drosophila melanogaster* recovering from a so-called transient sublethal heat shock, during which body temperature was increased ~5°C above normal body core temperature. Such a mild heat shock elicited a heat·shock response, characterized by the synthesis of new heat shock proteins normally almost absent in tissues of adult animals and by an increased synthesis of constitutively present or cognate heat shock proteins.

This event was followed by a transient increased tolerance to high, normally lethal temperatures (thermotolerance). Later it was found that not only the tolerance to enhanced temperature increases, but also the resistance toward other events like hypoxia, ischemia, inflammation, and exposure to such cellular toxins as heavy metals, endotoxins, and reactive oxygen species (cross-tolerance), all imposing serious stress upon tissues and their composing cells.

The increased resistance toward stressful events has been described in a great variety of organisms, organs, and tissues, including the heart.

During the last years, more and more information has become available on the specific role of individual heat shock proteins and on the mechanisms of transient activation of gene transcription, leading to the enhanced cardiac synthesis of these proteins.

A better insight has also been acquired in such subcellular targets as molecules or organelles that can be protected by heat shock proteins.

Now the phase is reached that upregulation of heat shock protein synthesis is considered as a powerful physiological, endogenous route for protecting crucial cellular homeostatic mechanisms against disturbing external factors.

Therefore, it is propagated that they provide a new therapeutic tool to protect the human heart during and after transient ischemic attacks.

Over the last 5 years comprehensive reviews have been published on the chaperone activity of heat shock proteins and on their role in specific tissues and pathological processes. The role of such individual heat shock proteins as the 70-kDa heat shock protein, small heat shock proteins and the heat shock factors has also been highlighted.

This survey concentrates on the transient protection of the cardiovascular system in various mammalian species when heat shock protein synthesis is upregulated.

The present state of the art regarding the stimuli, leading to the heat shock response, the transduction pathways involved in the activation transcription of the various heat shock genes, and the cellular structures protected by heat shock proteins is presented.

Such new experimental tools as transgenic animals and transfection techniques to enhance or inhibit the transcription of heat shock genes are discussed.

Although the physiological role of heat shock proteins and their protective potential under pathophysiological circumstances have been studied in detail in animal models, knowledge of the role and effects of these proteins in humans is still limited. Where possible the relevance of the observations for the human situation is discussed.

METALLOCHAPERONES, AN INTRACELLULAR SHUTTLE SERVICE FOR METAL IONS

Enzymes that employ transition metals as co-factors are housed in a wide variety of intracellular locations or are exported to the extracellular milieu.

A key question then arises: how are specific metal co-factors transported to diverse locations and subsequently sorted into the correct metalloenzymes?

The mechanisms by which these tasks are accomplished are just now being unraveled. A new family of soluble metal receptor proteins known as "metallochaperones" is emerging that act in the intracellular trafficking of metal ions.

Although transition elements can be quite toxic, these metal receptors are not detoxification proteins; they clearly function in a "chaperone-like" manner, guiding and protecting the metal ion while facilitating appropriate partnerships.

Here we will review the most recent advances in our understanding of copper metallochaperones and discuss mechanisms that may be relevant to other essential, yet potentially toxic, metal ions.

Conjugated Proteins

Covalently bonded to prosthetic groups such as glycoprotein and metalloprotein.

Cytokines

Regulate immunity, inflammation, apoptosis, and hematopoiesis.

Interleukins

Interleukins are the cytokines that act specifically as mediators between leucocytes.

The following table shows the major source and effects of various types of interleukins.

	Major source	Major effects
IL-1	Macrophages	Stimulation of T cells and antigen-presenting cells.B-cell growth and anibody production.Promotes hematopoiesis (blood cell formation).
IL-2	Activated T cells	Proliferation of activated T cells.
IL-3	T lymphocytes	Growth of blood cell precursors.
IL-4	T cells and mast cells	B-cell proliferation.IgE production.
IL-5	T cells and mast cells	Eosinophil growth.
IL-6	Activated T cells	Synergistic effects with IL-1 or TNFa.
IL-7	thymus and bone marrow stromal cells	Development of T cell and B cell precursors.
IL-8	Macrophages	Chemoattracts neutrophils.
IL-9	Activated T cells	Promotes growth of T cells and mast cells.
IL-10	Activated T cells, B cells and monocytes	Inhibits inflammatory and immune responses.
IL-11	Stromal cells	Synergistic effects on hematopoiesis.
IL-12	Macrophages, B cells	Promotes T_H1 cells while suppressing T_H2 functions
IL-13	T_H2 cells	Similar to IL-4 effects
IL-15	Epithelial cells and monocytes	Similar to IL-2 effects
IL-16	CD8 T cells	Chemoattracts CD4 T cells.
IL-17	Activated memory T cells	Promotes T cell proliferation.
IL-18	Macrophages	Induces IFNg production.

Most cytokines use the Jak-STAT pathway to transmit signals, but IL-1 uses a much more complex pathway.

Interferons

Interferons are the cytokines that can "interfere" with viral growth. They also have the ability to inhibit proliferation and modulate immune responses. Four types of interferons have been identified: IFN-α, IFN-β, IFN-ω and IFN-γ. The first three are Type I IFNs which have relatively high antiviral potency. IFN-γ is the Type II IFN, also called immune IFN.

Type I IFNs are produced by macrophages, neutrophils and other somatic cells in response to infection by viruses or bacteria. After they are released, they may bind to their receptors that are expressed on most cell types, resulting in the production of over 30 different proteins in the target cell.

Among them, two enzymes play a critical role in the inhibition of viral replication: RNA-dependent protein kinase (PKR) and 2'-5' oligoadenylate synthetase (2-5A synthetase). Both enzymes can be activated by double-stranded RNA which may be present in some viruses. The activated PKR can phosphorylate a protein (eIF2) to inhibit protein synthesis. The 2-5A synthetase produces oligoadenylate that can bind and activate a cellular endonuclease to degrade mRNA.

IFN-γ is produced in activated T_H1 and NK cells, particularly in response to IL-2 and IL-12. Its production is suppressed by IL-4, IL-10, and TGF-β. Binding of IFN-γ to its receptor increases the expression of class I MHC on all somatic cells. It also enhances the expression of class II MHC on antigen-presenting cells. IFN-γ may also activate macrophages, neutrophils and NK cells.

Tumor Necrosis Factors

Tumor necrosis factors are the cytokines produced mainly by macrophages and T lymphocytes that help regulate the immune response and hematopoiesis (blood cell formation). There are two types of TNF:

TNFα — also called cachectin, produced by macrophages.

TNFβ — also called lymphotoxin, produced by activated CD4+ T cells.

Human TNFα contains 157 amino acids, generated from a precursor protein of 233 amino acids by a converting enzyme called TACE. TNFβ contains 171 amino acids. Both TNFa and TNFb bind to the Type I and Type II receptors of the TNF receptor superfamily. The Type I receptor (TNFRI) has a molecular weight of 55kD while the Type II receptor (TNFRII) has 75kD. TNFRI contains a death domain which is involved in apoptosis. Most cytokines use the Jak-STAT pathway to transmit signals, but TNF may bind TNFRI to activate the NF-kB pathway.

TNF and IL-1 are structurally unrelated and bind different receptors, but their biological effects are quite similar. Both can stimulate lymphocyte proliferation and induce acute-phase responses.

Chemokines

Chemokines are the cytokines that may activate or chemoattract leukocytes. Each chemokine contains 65 ~ 120 amino acids, with molecular weight of 8 ~ 10 kD. Their receptors belong to G-protein-coupled receptors. Since the entry of HIV into host cells requires chemokine receptors, their antagonists are being developed to treat AIDS.

Hormones

Examples: insulin, growth hormone, and prolactin.

Prions

Prions are pathogenic variants of proteins that are naturally produced in nerve cells and certain other cells. The normal "healthy" prions are referred to as PrPc (Prion Protein cellular). The word "prion" stands for "proteinaceous infectious particle" and so should properly only be applied to the pathogenic variants. In this picture, the production of PrPc is illustrated from the nucleus at bottom right. RNA that codes for PrPc is produced in the nucleus and exits via the nuclear pore. The RNA then passes along ribosomes attached to the rER. PrPc is formed in the rER and then progresses up through the Golgi. At the upper face of the Golgi, vesicles containing PrPc bud off and travel to the cell surface. Here, they fuse with the cell membrane and so discharge their cargo at "PrPc". By this means, the cellular proteins come to sit on the exterior of the cell.

PrPc encounter rogue (purple) prions. These are termed PrPsc (sc stands for scrapie, the prion disease of sheep). The rogue prions seem to force the normal proteins to change shape. Both types of protein, the PrPc and their corresponding prions, are the same chemical - just different shapes. It is this shape-flipping which is equivalent to the transmission of infection.

Such a conformational shift or flip could happen at the cell surface or in caveolae (one is shown as a small invagination in the cell membrane).

Residual PrPc might continue to be flipped by contact with the rogue confomations for some time in these vesicles. Prions polymerise, finally appearing as purple fibrils in the picture at "P". PrPsc is resistant to degradation by the enzymes contained in the lysosomes that are seen floating nearby.

Consequently, PrPsc accumulates in the cell. PrPsc vesicles may also travel to the Golgi and intercept PrPc that is being processed there. In this way, PrPc particles could be switched to the rogue form before they reach the surface of the cell. By such mechanisms, PrPc might be switched to PrPsc at various points in and on the cell.

Prions could enter the brain along the axons of neurons. This probably happens by a retrograde flow of prion filled vesicles.

These are shown in the picture as purple spheres ascending the axon like elevators going up a shaft. Another route of entry could be the blood, probably in immune cells. A lymphocyte is shown exiting the capillary at bottom left where it could then contact the astrocyte. Astrocytes and other glial cells also support the production of prions.

Prions are implicated in BSE (Bovine Spongiform Encephalopathy or Mad Cow Disease) and its human counterpart nvCJD (new variant Creutzfeldt Jakob Disease). These and similar diseases are known as TSEs (Transmissible Spongiform Encephalopathies). TSEs afflict other species, most famously sheep (scrapie).

Structural Proteins

Collagen is the major insoluble fibrous protein in the extracellular matrix and in connective tissue. In fact, it is the single most abundant protein in the animal kingdom. There are at least 16 types of collagen, but 80 – 90 per cent of the collagen in the body consists of types I, II, and III. These collagen molecules pack together to form long thin *fibrils* of similar structure.

Type IV, in contrast, forms a two-dimensional reticulum; several other types associate with fibril-type collagens, linking them to each other or to other matrix components.

ZAt one time it was thought that all collagens were secreted by fibroblasts in connective tissue, but we now know that numerous epithelial cells make certain types of collagens. The various collagens and the structures they form all serve the same purpose, to help tissues withstand stretching.

Myosins are a large family of motor proteins found in eukaryotic tissues. They are responsible for actin-based motility. Most myosin molecules are composed of both a head and a tail domain.

- The *head domain* binds the filamentous actin, and uses ATP hydrolysis to generate force and to "walk" along the filament towards the (+) end (with the exception of one family member, myosin VI, which moves towards the (–) end).
- The *tail domain* generally mediates interaction with cargo molecules and/or other myosin subunits.

Myosin I

Myosin I's function is unknown, but it is believed to be responsible for vessicle transport or the contraction vacuole of cells.

Myosin II

Myosin II, responsible for skeletal muscle contraction, is perhaps the best-studied example of these properties.

- Myosin II contains two *heavy chains*, each about 2000 amino acids in length, which constitute the head and tail domains. Each of these heavy chains contains the N-terminal head domain, while the C-terminal tails take on a coiled-coil morphology, holding the two heavy chains together (imagine two snakes wrapped around each other, such as in a caduceus). Thus, myosin II has two heads.

- It also contains 4 *light chains* (2 per head), which bind the heavy chains in the "neck" region between the head and tail.

In muscle cells, it is myosin II that is responsible for producing the contractile force. Here, the long coiled-coil tails of the individual myosin molecules join together, forming the thick filaments of the sarcomere. The force-producing head domains stick out from the side of the thick filament, ready to walk along the adjacent actin-based thin filaments in response to the proper chemical signals.

Ubiquitin

The marker for protein degradation.If a protein binds to ubiquitin, it will be degraded by proteasome.

Like all macromolecular components of an organism, the proteome is in a dynamic state of synthesis and degradation. During proteolysis, the peptide bonds that link amino acids are hydrolyzed, and free amino acids are released.

The process is carried out by a diverse group of enzymes termed proteases. During proteolysis, the energy invested in the synthesis of the peptide bond is released. Distinct proteolytic mechanisms serve different physiological requirements and allow the organism to accommodate to changing environmental and pathophysiological conditions.

One should distinguish between destruction of "foreign" and "self" proteins. Foreign dietary proteins are degraded "outside" the body, in the lumen of the gastrointestinal tract. To avoid triggering an immune response, the epithelial lining of the digestive tract does not allow absorption of intact proteins into the body, and they are degraded to nonantigenic amino acids that are absorbed by the body and serve as building blocks for synthesis of its own proteins.

Self proteins can also be classified into extracellular and intracellular; the two groups of proteins are degraded via two distinct mechanisms. Extracellular proteins such as the blood coagulation factors, immunoglobulins, albumin, cargo-carrying proteins [such as the core protein of the low-density lipoprotein (LDL)], and peptide hormones (such as insulin) are taken up via pinocytosis or receptor-mediated endocytosis.

They are then carried via a series of vesicles (endosomes) that fuse with primary lysosomes where they are degraded. During this process, the extracellular proteins are never exposed to the intracellular environment (the cytosol) and remain "extracellular" (topologically) throughout. Degradation of proteins in lysosomes is not specific, and all engulfed proteins exposed to lysosomal proteases are degraded at approximately the same rate.

Several observations lead to the prediction that degradation of intracellular proteins must be carried out by completely distinct mechanisms. The process is highly specific, and different proteins have half-life times that vary from a few minutes to several days (e.g., the muscle proteins actin and myosin) and up to a few years (crystalline).

Furthermore, inhibitors of lysosomal degradation, weak bases such as chloroquine, for example, do not have any effect on degradation of intracellular proteins under basal metabolic conditions. These compounds titrate the normal acidic intralysosomal pH and bring it to a point that does not allow activity of the lysosomal proteases.

These findings led to the hypothesis that degradation of intracellular proteins must be carried out by a nonlysosomal proteolytic system that is endowed with a high degree of specificity toward its substrates. Also, the fact that the proteolytic enzymes and their substrates reside in the same cellular compartment predicted a requirement for tightly regulated machinery that uses metabolic energy for control. The discovery of the ubiquitin-proteasome proteolytic pathway has resolved these enigmas.

Degradation of a protein via the ubiquitin-proteasome pathway involves two discrete and successive steps:

- Tagging of the substrate by covalent attachment of multiple ubiquitin molecules and
- Degradation of the tagged protein by the 26S proteasome complex with release of free and reusable ubiquitin. This last process is mediated by ubiquitin recycling enzymes [deubiquitinating enzymes (DUBs)].

Conjugation of ubiquitin, a highly evolutionarily conserved 76-residue polypeptide, to the protein substrate proceeds via a three-step cascade mechanism. Initially, the ubiquitin-activating enzyme E1 activates ubiquitin in an ATP-requiring reaction to generate a high-energy thiol ester intermediate, E1-S~ubiquitin.

One of several E2 enzymes [ubiquitin-carrier proteins or ubiquitin-conjugating enzymes (UBCs)] transfers the activated ubiquitin moiety from E1, via an additional high-energy thiol ester intermediate, E2-S~ubiquitin, to the substrate that is specifically bound to a member of the ubiquitin-protein ligase family, E3.

There are a number of different classes of E3 enzymes. For the HECT (homologous to the E6-AP COOH terminus) domain E3s, the ubiquitin is transferred once again from the E2 enzyme to an active site Cys residue on the E3, to generate a third high-energy thiol ester intermediate, ubiquitin~ S-E3, before its transfer to the ligase-bound substrate. RING finger-containing E3s catalyze direct transfer of the activated ubiquitin moiety to the E3-bound substrate.

E3s catalyze the last step in the conjugation process: covalent attachment of ubiquitin to the substrate. The ubiquitin molecule is generally transferred to an -NH_2 group of an internal Lys residue in the substrate to generate a covalent isopeptide bond. In some cases, however, ubiquitin is conjugated to the NH_2-terminal amino group of the substrate. By successively adding activated ubiquitin moieties to internal Lys residues on the previously conjugated ubiquitin molecule, a polyubiquitin chain is synthesized.

The chain is recognized by the downstream 26S proteasome complex. Thus E3s play a key role in the ubiquitin-mediated proteolytic cascade since they serve as the specific recognition factors of the system. The proteasome is a large, 26S, multicatalytic protease that degrades polyubiquitinated proteins to small peptides. It is composed of two subcomplexes: a 20S core particle (CP) that carries the catalytic activity and a regulatory 19S regulatory particle (RP). The 20S CP is a barrel-shaped structure composed of four stacked rings, two identical outerrings and two identical innerrings.

The eukaryotic and rings are composed each of seven distinct subunits, giving the 20S complex the general structure. The catalytic sites are localized to some of the subunits. Each extremity of the 20S barrel can be capped by a 19S RP. One important function of the 19S RP is to recognize ubiquitinated proteins and other potential substrates of the proteasome.

A ubiquitin-binding subunit of the 19S RP has indeed been identified; however, its importance and mode of action have not been discerned. A second function of the 19S RP is to open an orifice in the ring that will allow entry of the substrate into the proteolytic chamber. Also, because a folded protein would not be able to fit through the narrow proteasomal channel, it is assumed that the 19S particle unfolds substrates and inserts them into the 20S CP.

Both the channel opening function and the unfolding of the substrate require metabolic energy, and indeed, the 19S RP contains six different ATPase subunits. After degradation of the substrate, short peptides derived from the substrate are released, as well as reusable ubiquitin.

A major unresolved question is, How does the system achieve its high specificity and selectivity? Why are certain proteins extremely stable in the cell, whereas others are extremely short-lived? Why are certain proteins degraded only at a particular time point during the cell cycle or only after specific extracellular stimuli, yet they are stable under most other conditions?

It appears that specificity of the ubiquitin system is determined by two distinct and unrelated groups of proteins:

E3s and ancillary proteins. First, within the ubiquitin system, substrates must be specifically recognized by an appropriate E3 as a prerequisite to their ubiquitination. In most cases however, substrates are not recognized in a constitutive manner and are not recognized directly by the E3. In some instances, the E3 must "be switched on" by undergoing posttranslational modification to yield an active form that recognizes the substrate. In many other cases, it is the substrate that must undergo a certain change that renders it susceptible for recognition.

The stability of additional proteins depends on association with ancillary proteins such as molecular chaperones that act as recognition elements in *trans* and serve as a link to the appropriate ligase. Others, such as certain transcription factors, have to dissociate from the specific DNA sequence to which they bind in order to be recognized by the system.

Stability of yet other proteins depends on oligomerization. Thus, in addition to the E3s themselves, modifying enzymes (such as kinases), ancillary proteins, or DNA sequences to which substrates bind, also play an important role in the recognition process.

Ubiquitin-mediated proteolysis of a variety of cellular proteins plays an important role in many basic cellular processes. Among these are regulation of cell cycle and division, differentiation and development, involvement in the cellular response to stress and extracellular effectors, morphogenesis of neuronal networks, modulation of cell surface receptors, ion channels and the secretory pathway, DNA repair, transcriptional regulation, transcriptional silencing, long-term memory, circadian rhythms, regulation of the immune and inflammatory responses, and biogenesis of organelles.

The list of cellular proteins that are targeted by ubiquitin is growing rapidly. Among them are cell cycle regulators such as cyclins, cyclin-dependent kinase inhibitors, and proteins involved in sister chromatid separation, tumor suppressors, as well as transcriptional activators and their inhibitors.

Cell surface receptors and endoplasmic reticulum (ER) proteins are also targeted by the system. Finally, mutated and denatured/misfolded proteins are recognized specifically and are removed efficiently. In this capacity, the system is a key player in the cellular quality control and defence mechanisms. With the numerous substrate proteins targeted and the multitude of processes involved, it is not surprising that aberrations in the ubiquitin system have been implicated in the pathogenesis of many inherited and acquired human pathologies. In some cases, the linkage between the proteolytic system and the resulting pathology is direct, whereas in others it is less obvious.

It is impossible to cover in a single review, or even in a comprehensive monograph, all that we currently know of the ubiquitin system, its enzymatic components, ancillary proteins, modes of action, mechanisms of substrate recognition, modes of regulation, and most of all, the numerous processes it is involved in or the consequences of aberrations in its activity.

We therefore decided to bring to the reader an updated view of the general components of the system, the conjugating enzymes and the proteasome, explain their mode of action, and highlight, via specific examples, several of the processes in which the system is involved.

We will end by describing in greater detail the recently evolving field of the involvement of this system in pathogenesis of human diseases. For more details about the ubiquitin system, the reader is referred to the many specific reviews written on it recently.

Chapter 3

Carbohydrates as Biomolecule

Carbohydrates are the most abundant class of organic compounds found in living organisms. They originate as products of photosynthesis, an endothermic reductive condensation of carbon dioxide requiring light energy and the pigment chlorophyll.

$$CO_2 + n\,H_2O + \text{energy} \rightarrow C_n(H_{2n}O_n) + n\,O_2$$

Complexity	**Simple Carbohydrates**		**Complex**		
Carbohydrates	monosaccharides		disaccharides,		
			oligosaccharides & polysaccharides		
Size	**Tetrose**	**Pentose**	**Hexose**	**Heptose**	
	C_4 sugars	C_5 sugars	C_6 sugars	C_7 sugars	etc.

C=O	**Aldose**
Function	sugars having an aldehyde function or an acetal equivalent.
	Ketose
	sugars having a ketone function or an acetal equivalent.
Reactivity	**Reducing** sugars oxidized by Tollens' reagent (or Benedict's or Fehling's reagents).
	Non-reducing
	sugars not oxidized by Tollens' or other reagents.

As noted here, the formulas of many carbohydrates can be written as carbon hydrates, $C_n(H_2O)_n$, hence their name. The carbohydrates are a major source of metabolic energy, both for plants and for animals that depend on plants for food. Aside from the sugars and starches that meet this vital nutritional role, carbohydrates also serve as a structural material (cellulose), a component of the energy transport compound ATP, recognition sites on cell surfaces and one of three essential components of DNA and RNA.

Carbohydrates are called saccharides or, if they are relatively small, sugars. Several classifications of carbohydrates have proven useful and are outlined in the above table.

GLUCOSE

Carbohydrates have been given non-systematic names, although the suffix ose is generally used. The most common carbohydrate is glucose ($C_6H_{12}O_6$). Glucose is a monosaccharide, an aldohexose and a reducing sugar. The general structure of glucose and many other aldohexoses was established by simple chemical reactions. The following diagram illustrates the kind of evidence considered, although some of the reagents shown here are different from those used by the original scientists.

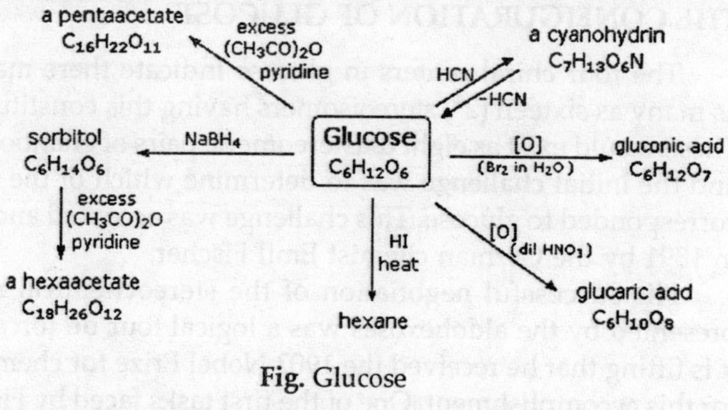

Fig. Glucose

Hot hydriodic acid (HI) was often used to reductively remove oxygen functional groups from a molecule and in the case of glucose this treatment gave hexane (in low yield). From this it was concluded that the six carbons are in an unbranched chain. The presence of an aldehyde carbonyl group was deduced from cyanohydrin formation, its reduction to the hexa-alcohol sorbitol, also called glucitol and mild oxidation to the mono-carboxylic acid, glucuronic acid.

Somewhat stronger oxidation by dilute nitric acid gave the diacid, glucaric acid, supporting the proposal of a six-carbon chain. The five oxygens remaining in glucose after the aldehyde was accounted for were thought to be in hydroxyl groups, since a penta-acetate derivative could be made. These hydroxyl groups were assigned, one each, to the last five carbon atoms, because geminal hydroxyl groups are normally unstable relative to the carbonyl compound formed by loss of water. The four middle carbon atoms in the glucose chain are centers of chirality and are colored red.

Glucose and other saccharides are extensively cleaved by periodic acid, thanks to the abundance of vicinal diol moieties in their structure. This oxidative cleavage, known as the Malaprade reaction is particularly useful for the analysis of selective O-substituted derivatives of saccharides, since ether functions do not react. The stoichiometry of aldohexose cleavage is shown in the following equation.

$$HOCH_2(CHOH)_4CHO + 5\,HIO_4 \rightarrow H_2C$$
$$= O + 5\,HCO_2H + 5\,HIO_3$$

THE CONFIGURATION OF GLUCOSE

The four chiral centers in glucose indicate there may be as many as sixteen (2^4) stereoisomers having this constitution. These would exist as eight diastereomeric pairs of enantiomers and the initial challenge was to determine which of the eight corresponded to glucose. This challenge was accepted and met in 1891 by the German chemist Emil Fischer.

His successful negotiation of the stereochemical maze presented by the aldohexoses was a logical tour de force and it is fitting that he received the 1902 Nobel Prize for chemistry for this accomplishment. One of the first tasks faced by Fischer was to devise a method of representing the configuration of each chiral cenre in an unambiguous manner. To this end, he invented a simple technique for drawing chains of chiral centers, that we now call the Fischer projection formula. At the time Fischer undertook the glucose project it was not possible to establish the absolute configuration of an enantiomer.

Consequently, Fischer made an arbitrary choice for (+)-glucose and established a network of related aldose configurations that he called the D-family. The mirror images of these configurations were then designated the L-family of aldoses. To illustrate using present day knowledge, Fischer projection formulas and names for the D-aldose family (three to six-carbon atoms) are shown below, with the asymmetric carbon atoms (chiral centers) colored red.

The last chiral cenre in an aldose chain (farthest from the aldehyde group) was chosen by Fischer as the D/L designator site. If the hydroxyl group in the projection formula pointed to the right, it was defined as a member of the D-family. A left directed hydroxyl group then represented the L-family.

Fischer's initial assignment of the D-configuration had a 50:50 chance of being right, but all his subsequent conclusions concerning the relative configurations of various aldoses were soundly based. In 1951 x-ray fluorescence studies of (+)-tartaric acid, carried out in the Netherlands by Johannes Martin Bijvoet (pronounced "buy foot"), proved that Fischer's choice was correct.

It is important to recognize that the sign of a compound's specific rotation (an experimental number) does not correlate with its configuration (D or L). It is a simple matter to measure an optical rotation with a polarimeter. Determining an absolute configuration usually requires chemical interconversion with known compounds by stereospecific reaction paths.

OXIDATION

Sugars may be classified as reducing or non-reducing based on their reactivity with Tollens', Benedict's or Fehling's reagents. If a sugar is oxidized by these reagents it is called reducing, since the oxidant (Ag^{+} or Cu^{+2}) is reduced in the reaction, as evidenced by formation of a silver mirror or precipitation of cuprous oxide. The Tollens' test is commonly used to detect aldehyde functions; and because of the facile interconversion of ketoses and aldoses under the basic conditions of this test, ketoses such as fructose also react and are classified as reducing sugars. When the aldehyde function of an aldose is oxidized to a carboxylic acid the product is called an aldonic acid. Because of the 2º hydroxyl functions that are also present in these compounds, a mild oxidizing agent such as hypobromite must be used for this conversion.

If both ends of an aldose chain are oxidized to carboxylic acids the product is called an aldaric acid. By converting an aldose to its corresponding aldaric acid derivative, the ends of the chain become identical (this could also be accomplished by reducing the aldehyde to CH_2OH, as noted below). Such an operation will disclose any latent symmetry in the remaining molecule.

Thus, ribose, xylose, allose and galactose yield achiral aldaric acids which are, of course, not optically active. Other aldose sugars may give identical chiral aldaric acid products, implying a unique configurational relationship. Remember, a Fischer projection formula may be rotated by 180º in the plane of projection without changing its configuration.

REDUCTION

Sodium borohydride reduction of an aldose makes the ends of the resulting alditol chain identical, $HOCH_2(CHOH)_nCH_2OH$, thereby accomplishing the same configurational change produced by oxidation to an aldaric acid. Thus, allitol and galactitol from reduction of allose and galactose are achiral and altrose and talose are reduced to the same chiral alditol. A summary of these redox reactions and derivative nomenclature is given in the following table.

Derivatives of $HOCH_2(CHOH)_nCHO$

HOBr Oxidation → $HOCH_2(CHOH)_nCO_2H$ an Aldonic Acid

HNO_3 Oxidation → $H_2OC(CHOH)_nCO_2H$ an Aldaric Acid

$NaBH_4$ Reduction → $HOCH_2(CHOH)_nCH_2OH$ an Alditol

OSAZONE FORMATION

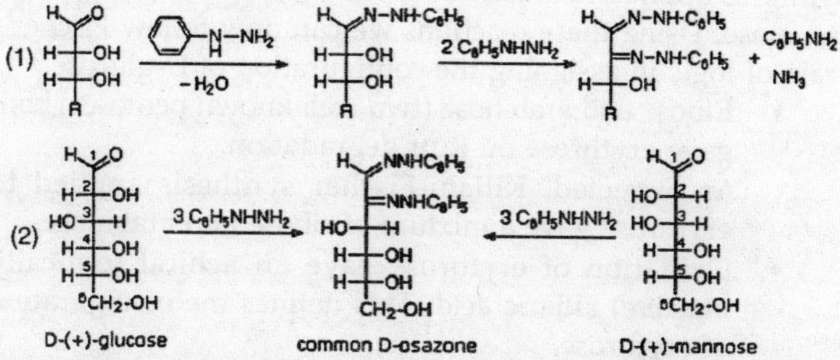

The osazone reaction was developed and used by Emil Fischer to identify aldose sugars differing in configuration only at the alpha-carbon. The upper equation shows the general form of the osazone reaction, which effects an alpha-carbon oxidation with formation of a bis-phenylhydrazone, known as an osazone. Application of the osazone reaction to D-glucose and D-mannose demonstrates that these compounds differ in configuration only at C-2.

CHAIN SHORTENING AND LENGTHENING

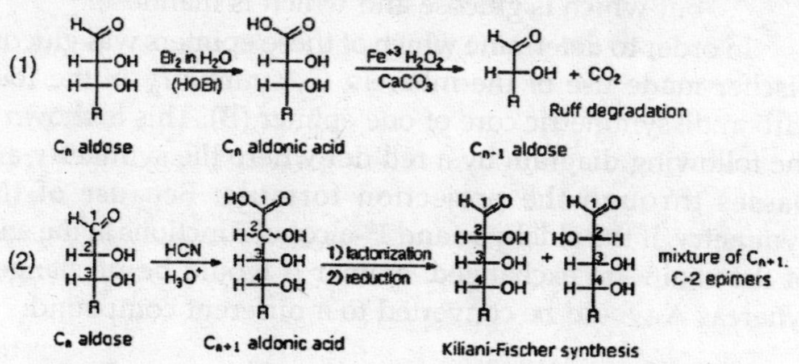

These two procedures permit an aldose of a given size to be related to homologous smaller and larger aldoses. The importance of these relationships may be seen in the array of aldose structures presented earlier, where the structural connections are given by the dashed blue lines. Thus Ruff degradation of the pentose arabinose gives the tetrose erythrose. Working in the opposite direction, a Kiliani-Fischer synthesis applied to arabinose gives a mixture of glucose and mannose. Using these reactions we can now follow Fischer's train of logic in assigning the configuration of D-glucose.

- Ribose and arabinose (two well known pentoses) both gave erythrose on Ruff degradation.

  As expected, Kiliani-Fischer synthesis applied to erythrose gave a mixture of ribose and arabinose.
- Oxidation of erythrose gave an achiral (optically inactive) aldaric acid. This defines the configuration of erythrose.
- Oxidation of ribose gave an achiral (optically inactive) aldaric acid. This defines the configuration of both ribose and arabinose.
- Ruff shortening of glucose gave arabinose and Kiliani-Fischer synthesis applied to arabinose gave a mixture of glucose and mannose.
- Glucose and mannose are therefore epimers at C-2, a fact confirmed by the common product from their osazone reactions.
- A pair of structures for these epimers can be written, but which is glucose and which is mannose?

In order to determine which of these epimers was glucose, Fischer made use of the inherent C_2 symmetry in the four-carbon dissymmetric core of one epimer (B). This is shown in the following diagram by a red dot where the symmetry axis passes through the projection formula. Because of this symmetry, if the aldehyde and 1º-alcohol functions at the ends of the chain are exchanged, epimer B would be unchanged; whereas A would be converted to a different compound.

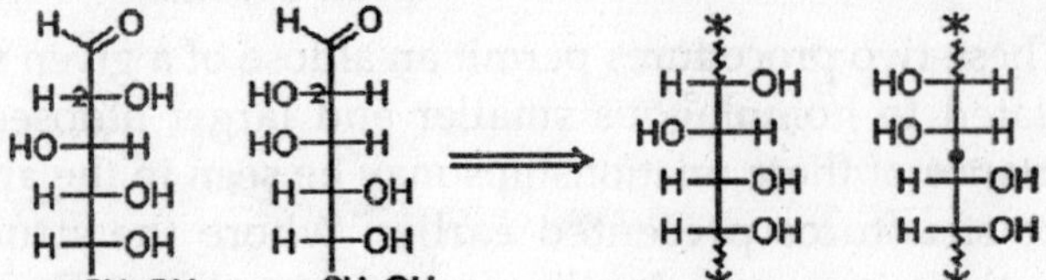

Fischer looked for and discovered a second aldohexose that represented the end group exchange for the epimer lacking the latent C_2 symmetry (A). This compound was L-(+)-gulose and its exchange relationship to D-(+)-glucose was demonstrated be oxidation to a common aldaric acid product.

KETOSES

If a monosaccharide has a carbonyl function on one of the inner atoms of the carbon chain it is classified as a ketose. Dihydroxyacetone may not be a sugar, but it is included as the ketose analog of glyceraldehyde.

The carbonyl group is commonly found at C-2, as illustrated by the following examples (chiral centers are colored red). As expected, the carbonyl function of a ketose may be reduced by sodium borohydride, usually to a mixture of epimeric products. D-Fructose, the sweetest of the common natural sugars, is for example reduced to a mixture of D-glucitol (sorbitol) and D-mannitol, named after the aldohexoses from which they may also be obtained by analogous reduction. Mannitol is itself a common natural carbohydrate.

Although the ketoses are distinct isomers of the aldose monosaccharides, the chemistry of both classes is linked due to their facile interconversion in the presence of acid or base catalysts. This interconversion and the corresponding epimerization at sites alpha to the carbonyl functions, occurs by way of an enediol tautomeric intermediate.

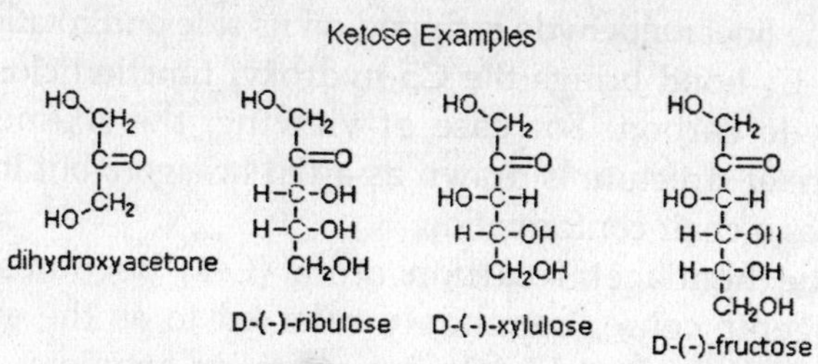

Because of base-catalyzed isomerizations of this kind, the Tollens' reagent is not useful for distinguishing aldoses from ketoses or for specific oxidation of aldoses to the corresponding aldonic acids. Oxidation by HOBr is preferred for the latter conversion.

ANOMERIC FORMS OF GLUCOSE

Fischers brilliant elucidation of the configuration of glucose did not remove all uncertainty concerning its structure. Two different crystalline forms of glucose were reported in 1895. Each of these gave all the characteristic reactions of glucose and when dissolved in water equilibrated to the same mixture. This equilibration takes place over a period of many minutes and the change in optical activity that occurs is called mutarotation. These facts are summarized in the diagram below.

When glucose was converted to its pentamethyl ether (reaction with excess CH_3I & AgOH), two different isomers were isolated and neither exhibited the expected aldehyde reactions. Acid-catalyzed hydrolysis of the pentamethyl ether derivatives, however, gave a tetramethyl derivative that was oxidized by Tollen's reagent and reduced by sodium borohydride, as expected for an aldehyde.

The search for scientific truth often proceeds in stages and the structural elucidation of glucose serves as a good example. Somehow a new stereogenic cenre must be created and the aldehyde must be deactivated in the pentamethyl derivative. A simple solution to this dilemma is achieved by converting the open aldehyde structure for glucose into a cyclic hemiacetal, called a glucopyranose.

The linear aldehyde is tipped on its side and rotation about the C_4-C_5 bond brings the C5-hydroxyl function close to the aldehyde carbon. For ease of viewing, the six-membered hemiacetal structure is drawn as a flat hexagon, but it actually assumes a chair conformation.

The hemiacetal carbon atom (C-1) becomes a new stereogenic cenre, commonly referred to as the anomeric carbon and the á and â-isomers are called anomers.

First, we know that hemiacetals are in equilibrium with their carbonyl and alcohol components when in solution. Consequently, fresh solutions of either alpha or beta-glucose crystals in water should establish an equilibrium mixture of both anomers, plus the open chain chain form.

Second, a pentamethyl ether derivative of the pyranose structure converts the hemiacetal function to an acetal. Acetals are stable to base, so this product should not react with Tollen's reagent or be reduced by sodium borohydride. Acid hydrolysis of acetals regenerates the carbonyl and alcohol components and in the case of the glucose derivative this will be a tetramethyl ether of the pyranose hemiacetal. This compound will, of course, undergo typical aldehyde reactions.

CYCLIC FORMS OF MONOSACCHARIDES

The preferred structural form of many monosaccharides may be that of a cyclic hemiacetal. Five and six-membered rings are favored over other ring sizes because of their low angle and eclipsing strain.

Cyclic structures of this kind are termed furanose (five-membered) or pyranose (six-membered), reflecting the ring size relationship to the common heterocyclic compounds furan and pyran shown on the right. Ribose, an important aldopentose, commonly adopts a furanose structure by convention for the D-family, the five-membered furanose ring is drawn in an edgewise projection with the ring oxygen positioned away from the viewer.

The anomeric carbon atom (colored red here) is placed on the right. The upper bond to this carbon is defined as beta, the lower bond then is alpha. The cyclic pyranose forms of various monosaccharides are often drawn in a flat projection known as a Haworth formula, after the British chemist, Norman Haworth.

As with the furanose ring, the anomeric carbon is placed on the right with the ring oxygen to the back of the edgewise view. In the D-family, the alpha and beta bonds have the same orientation defined for the furanose ring. These Haworth formulas are convenient for displaying stereochemical relationships, but do not represent the true shape of the molecules. We know that these molecules are actually puckered in a fashion we call a chair conformation.

Examples of four typical pyranose structures are shown below, both as Haworth projections and as the more representative chair conformers. The anomeric carbons are colored red.

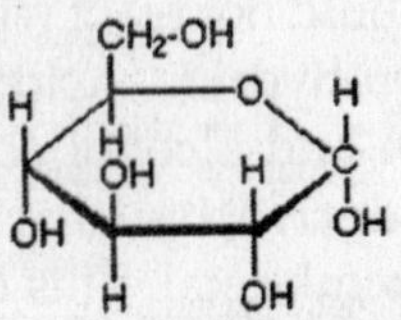

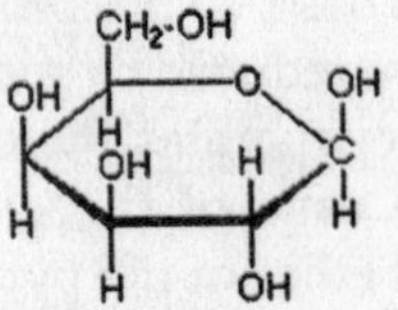

α-D-glucopyranose β-D-galactopyranose

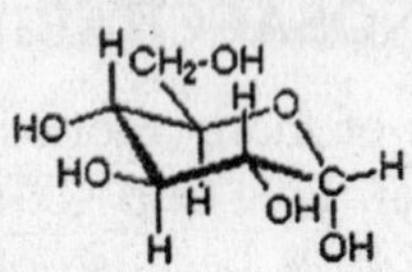

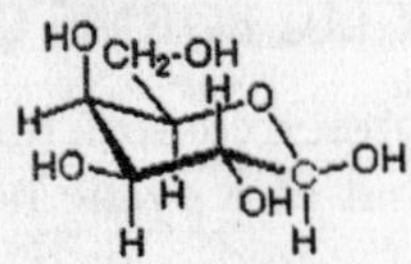

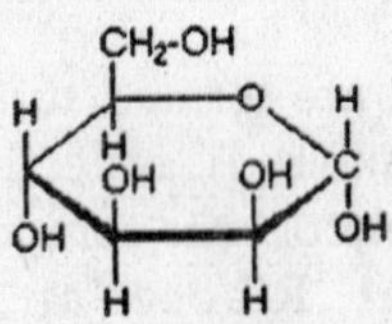

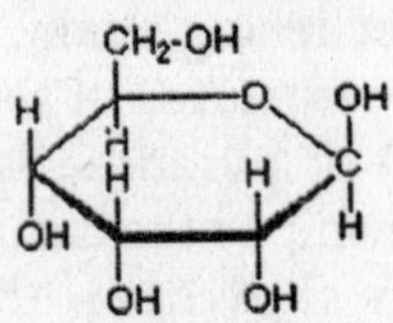

α-D-mannopyranose β-D-allopyranose

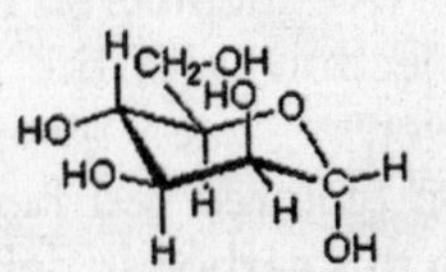

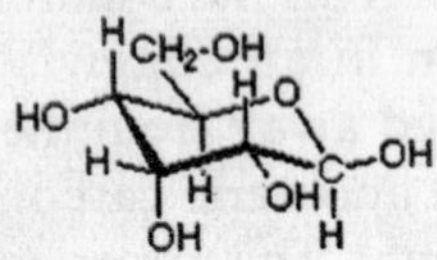

The size of the cyclic hemiacetal ring adopted by a given sugar is not constant, but may vary with substituents and other structural features. Aldolhexoses usually form pyranose rings and their pentosc homologs tend to prefer the furanose form, but there are many counter examples.

The formation of acetal derivatives illustrates how subtle changes may alter this selectivity. A pyranose structure for D-glucose is drawn in the rose-shaded box on the left. Acetal derivatives have been prepared by acid-catalyzed reactions with benzaldehyde and acetone. As a rule, benzaldehyde forms six-membered cyclic acetals, whereas acetone prefers to form five-membered acetals. The top equation shows the formation and some reactions of the 4,6-O-benzylidene acetal, a commonly employed protective group.

A methyl glycoside derivative of this compound leaves the C-2 and C-3 hydroxyl groups exposed to reactions such as the periodic acid cleavage, shown as the last step. The formation of an isopropylidene acetal at C-1 and C-2, cenre structure, leaves the C-3 hydroxyl as the only unprotected function. Selective oxidation to a ketone is then possible. Finally, direct di-O-isopropylidiene derivatization of glucose by reaction with excess acetone results in a change to a furanose structure in which the C-3 hydroxyl is again unprotected.

However, the same reaction with D-galactose, shown in the blue-shaded box, produces a pyranose product in which the C-6 hydroxyl is unprotected. Both derivatives do not react with Tollens' reagent. This difference in behaviour is attributed to the cis-orientation of the C-3 and C-4 hydroxyl groups in galactose, which permits formation of a less strained five-membered cyclic acetal, compared with the trans-C-3 and C-4 hydroxyl groups in glucose. Derivatizations of this kind permit selective reactions to be conducted at different locations in these highly functionalized molecules.

GLYCOSIDES

Acetal derivatives formed when a monosaccharide reacts with an alcohol in the presence of an acid catalyst are called glycosides. This reaction is illustrated for glucose and methanol in the diagram below. In naming of glycosides, the "oseΔ suffix of the sugar name is replaced by "osideΔ and the alcohol group name is placed first. As is generally true for most aldols, glycoside formation involves the loss of an equivalent of water. The diether product is stable to base and alkaline oxidants such as Tollen's reagent. Since acid-catalyzed aldolization is reversible, glycosides may be hydrolyzed back to their alcohol and sugar components by aqueous acid.

The anomeric methyl glucosides are formed in an equilibrium ratio of 66% alpha to 34% beta. In the case of glucose, the substituents on the beta-anomer are all equatorial, whereas the C-1 substituent in the alpha-anomer changes to axial. Since substituents on cyclohexane rings prefer an equatorial location over axial (methoxycyclohexane is 75% equatorial), the preference for alpha-glycopyranoside formation is unexpected and is referred to as the anomeric effect.

Glycosides abound in biological systems. By attaching a sugar moiety to a lipid or benzenoid structure, the solubility and other properties of the compound may be changed substantially.

Because of the important modifying influence of such derivatization, numerous enzyme systems, known as glycosidases, have evolved for the attachment and removal of sugars from alcohols, phenols and amines. Chemists refer to the sugar component of natural glycosides as the glycon and the alcohol component as the aglycon. Salicin, one of the oldest herbal remedies known, was the model for the synthetic analgesic aspirin.

A large class of hydroxylated, aromatic oxonium cations called anthocyanins provide the red, purple and blue colors of many flowers, fruits and some vegetables. Peonin is one example of this class of natural pigments, which exhibit a pronounced pH colour dependence. The oxonium moiety is only stable in acidic environments and the colour changes or disappears when base is added.

The complex changes that occur when wine is fermented and stored are in part associated with glycosides of anthocyanins. Finally, amino derivatives of ribose, such as cytidine play important roles in biological phosphorylating agents, coenzymes and information transport and storage materials.

DISACCHARIDES

When the alcohol component of a glycoside is provided by a hydroxyl function on another monosaccharide, the compound is called a disaccharide. Four examples of disaccharides composed of two glucose units are shown in the following diagram. The individual glucopyranose rings are labled A and B and the glycoside bonding is circled in light blue. Notice that the glycoside bond may be alpha, as in maltose and trehalose, or beta as in cellobiose and gentiobiose. Acid-catalyzed hydrolysis of these disaccharides yields glucose as the only product.

Enzyme-catalyzed hydrolysis is selective for a specific glycoside bond, so an alpha-glycosidase cleaves maltose and trehalose to glucose, but does not cleave cellobiose or gentiobiose. A beta-glycosidase has the opposite activity. In order to draw a representative structure for cellobiose, one of the glucopyranose rings must be rotated by 180º, but this feature is often omitted in favour of retaining the usual perspective for the individual rings.

The bonding between the glucopyranose rings in cellobiose and maltose is from the anomeric carbon in ring A to the C-4 hydroxyl group on ring B. This leaves the anomeric carbon in ring B free, so cellobiose and maltose both may assume alpha and beta anomers at that site. Gentiobiose has a beta-glycoside link, originating at C-1 in ring A and terminating at C-6 in ring B. Its alpha-anomer is drawn in the diagram.

Because cellobiose, maltose and gentiobiose are hemiacetals they are all reducing sugars (oxidized by Tollen's reagent). Trehalose, a disaccharide found in certain mushrooms, is a bis-acetal and is therefore a non-reducing sugar. A systematic nomenclature for disaccharides exists, but as the following examples illustrate, these are often lengthy.

Cellobiose: 4-O-β-D-Glucopyranosyl-D-glucose

Maltose: 4-O-α-D-Glucopyranosyl-D-glucose

Gentiobiose: 6-O-β-D-Glucopyranosyl-D-glucose

Trehalose: α-D-Glucopyranosyl-α-D-glucopyranoside

Although all the disaccharides shown here are made up of two glucopyranose rings, their properties differ in interesting ways. Maltose, sometimes called malt sugar, comes from the hydrolysis of starch. It is about one third as sweet as cane sugar (sucrose), is easily digested by humans and is fermented by yeast. Cellobiose is obtained by the hydrolysis of cellulose.

It has virtually no taste, is indigestible by humans and is not fermented by yeast. Some bacteria have beta-glucosidase enzymes that hydrolyze the glycosidic bonds in cellobiose and cellulose. The presence of such bacteria in the digestive tracts of cows and termites permits these animals to use cellulose as a food. Finally, it may be noted that trehalose has a distinctly sweet taste, but gentiobiose is bitter.

Disaccharides made up of other sugars are known, but glucose is often one of the components.. Lactose, also known as milk sugar, is a galactose-glucose compound joined as a beta-glycoside. It is a reducing sugar because of the hemiacetal function remaining in the glucose moiety. Many adults, particulaly those from regions where milk is not a dietary staple, have a metabolic intolerance for lactose. Infants have a digestive enzyme which cleaves the beta-glycoside bond in lactose, but production of this enzyme stops with weaning.

Cheese is less subject to the lactose intolerance problem, since most of the lactose is removed with the whey. Sucrose, or cane sugar, is our most commonly used sweetening agent. It is a non-reducing disaccharide composed of glucose and fructose joined at the anomeric carbon of each by glycoside bonds (one alpha and one beta). In the formula shown here the fructose ring has been rotated 180º from its conventional perspective.

POLYSACCHARIDES

As the name implies, polysaccharides are large high-molecular weight molecules constructed by joining monosaccharide units together by glycosidic bonds. They are sometimes called glycans. The most important compounds in this class, cellulose, starch and glycogen are all polymers of glucose.

This is easily demonstrated by acid-catalyzed hydrolysis to the monosaccharide. Since partial hydrolysis of cellulose gives varying amounts of cellobiose, we conclude the glucose units in this macromolecule are joined by beta-glycoside bonds between C-1 and C-4 sites of adjacent sugars. Partial hydrolysis of starch and glycogen produces the disaccharide maltose together with low molecular weight dextrans, polysaccharides in which glucose molecules are joined by alpha-glycoside links between C-1 and C-6, as well as the alpha C-1 to C-4 links found in maltose.

Over half of the total organic carbon in the earth's biosphere is in cellulose. Cotton fibres are essentially pure cellulose and the wood of bushes and trees is about 50% cellulose. As a polymer of glucose, cellulose has the formula $(C_6H_{10}O_5)n$ where n ranges from 500 to 5,000, depending on the source of the polymer.

The glucose units in cellulose are linked in a linear fashion, as shown in the drawing below. The beta-glycoside bonds permit these chains to stretch out and this conformation is stabilized by intramolecular hydrogen bonds. A parallel orientation of adjacent chains is also favored by intermolecular hydrogen bonds. Although an individual hydrogen bond is relatively weak, many such bonds acting together can impart great stability to certain conformations of large molecules.

Most animals cannot digest cellulose as a food and in the diets of humans this part of our vegetable intake functions as roughage and is eliminated largely unchanged. Some animals (the cow and termites, for example) harbour intestinal microorganisms that breakdown cellulose into monosacharide nutrients by the use of beta-glycosidase enzymes.

Cellulose is commonly accompanied by a lower molecular weight, branched, amorphous polymer called hemicellulose. In contrast to cellulose, hemicellulose is structurally weak and is easily hydrolyzed by dilute acid or base. Also, many enzymes catalyze its hydrolysis. Hemicelluloses are composed of many D-pentose sugars, with xylose being the major component. Mannose and mannuronic acid are often present, as well as galactose and galacturonic acid.

Starch is a polymer of glucose, found in roots, rhizomes, seeds, stems, tubers and corms of plants, as microscopic granules having characteristic shapes and sizes. Most animals, including humans, depend on these plant starches for nourishment. The structure of starch is more complex than that of cellulose. The intact granules are insoluble in cold water, but grinding or swelling them in warm water causes them to burst. The released starch consists of two fractions. About 20% is a water soluble material called amylose. Molecules of amylose are linear chains of several thousand glucose units joined by alpha C-1 to C-4 glycoside bonds.

Amylose solutions are actually dispersions of hydrated helical micelles. The majority of the starch is a much higher molecular weight substance, consisting of nearly a million glucose units and called amylopectin. Molecules of amylopectin are branched networks built from C-1 to C-4 and C-1 to C-6 glycoside links and are essentially water insoluble. The branching in this diagram is exaggerated, since on average, branches only occur every twenty five glucose units.

Hydrolysis of starch, usually by enzymatic reactions, produces a syrupy liquid consisting largely of glucose. When cornstarch is the feedstock, this product is known as corn syrup. It is widely used to soften texture, add volume, prohibit crystallization and enhance the flavour of foods.

Glycogen is the glucose storage polymer used by animals. It has a structure similar to amylopectin, but is even more highly branched (about every tenth glucose unit). The degree of branching in these polysaccharides may be measured by enzymatic or chemical analysis.

SYNTHETIC MODIFICATION OF CELLULOSE

Cotton, probably the most useful natural fibre, is nearly pure cellulose. The manufacturie of textiles from cotton involves physical manipulation of the raw material by carding, combing and spinning selected fibers. For fabrics the best cotton has long fibers and short fibers or cotton dust are removed. Crude cellulose is also available from wood pulp by dissolving the lignan matrix surrounding it. These less desirable cellulose sources are widely used for making paper.

In order to expand the ways in which cellulose can be put to practical use, chemists have devised techniques for preparing solutions of cellulose derivatives that can be spun into fibers, spread into a film or cast in various solid forms. A key factor in these transformations are the three free hydroxyl groups on each glucose unit in the cellulose chain, $-[C_6H_7O(OH)_3]_n$. Esterification of these functions leads to polymeric products having very different properties compared with cellulose itself.

Cellulose Nitrate, first prepared over 150 years ago by treating cellulose with nitric acid, is the earliest synthetic polymer to see general use. The fully nitrated compound, $-[C_6H_7O(ONO_2)_3]_n$, called guncotton, is explosively flammable and is a component of smokeless powder. Partially nitrated cellulose is called pyroxylin. Pyroxylin is soluble in ether and at one time was used for photographic film and lacquers. The high flammability of pyroxylin caused many tragic cinema fires during its period of use. Furthermore, slow hydrolysis of pyroxylin yields nitric acid, a process that contributes to the deterioration of early motion picture films in storage.

Cellulose Acetate, $-[C_6H_7O(OAc)_3]_n$, is less flammable than pyroxylin and has replaced it in most applications. It is prepared by reaction of cellulose with acetic anhydride and an acid catalyst.

The properties of the product vary with the degree of acetylation. Some chain shortening occurs unavoidably in the preparations. An acetone solution of cellulose acetate may be forced through a spinnerette to generate filaments, called acetate rayon, that can be woven into fabrics.

Viscose Rayon, is prepared by formation of an alkali soluble xanthate derivative that can be spun into a fibre that reforms the cellulose polymer by acid quenching. The following general equation illustrates these transformations. The product fibre is called viscose rayon.

$$\underset{\text{cellulose}}{ROH} \overset{NaOH}{\rightleftarrows} RO^{(-)}Na^{(+)} + S{=}C{=}S \rightarrow \underset{\text{viscose solution}}{RO\text{-}CS_2^{(-)}Na^{(+)}} \xrightarrow{H_3O^{(+)}} \underset{\text{rayon}}{ROH}$$

STRUCTURE OF CARBOHYDRATES

MONOSACCHARIDES STRUCTURES

Sugars can be defined as polyhydroxy aldehydes or ketones. Hence the simplest sugars contain at least three carbons. The most common are the aldo- and keto-trioses, tetroses, pentoses and hexoses. The simplest ^{3}C sugars are glyceraldehye and dihydroxyacetone.

Glucose, an aldo-hexose, is a central sugar in metabolism. It and other 5 and 6C sugars can cyclize through intramolecular nucleophilic attack of one of the OH's on the carbonyl C of the aldehyde or ketone.

Such intramolecular reactions occur if stable 5 or 6 member rings can form. The resulting rings are labeled furanose (5 member) or pyranose (6 member) based on their similarity to furan and pyran.

On nucleophilic attack to form the ring, the carbonyl O becomes an OH which points either below the ring (a anomer) or above the ring (β anomer).

Monosaccharides in solution exist as equilbrium mixtures of the straight and cyclic forms. In solution, glucose (Glc) is mostly in the pyranose form, fructose is 67% pyranose and 33% furanose and ribose is 75% furanose and 25% pyranose. However, in polysaccharides, Glc is exclusively pyranose and fructose and ribose are furanoses. Sugars can be drawn in the straight chain form as either Fisher projections or perspective structural formulas.

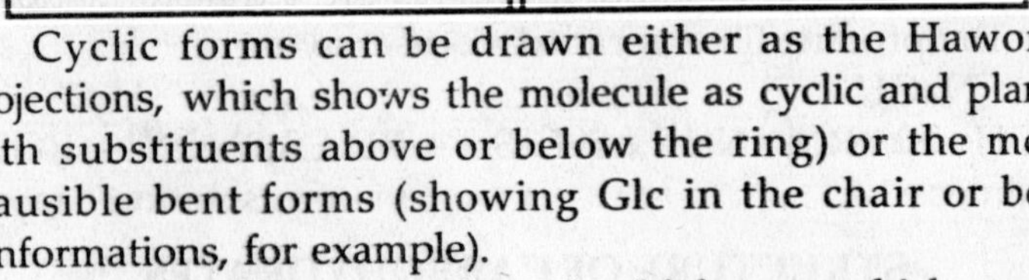

Cyclic forms can be drawn either as the Haworth projections, which shows the molecule as cyclic and planar with substituents above or below the ring) or the more plausible bent forms (showing Glc in the chair or boat conformations, for example).

b-D-glucopyranose is the only aldohexose which can be drawn with all its bulky substituents (OH and CH_2OH) in equatorial positions, which probably accounts for its widespread prevalence in nature.

Haworth projections are more realistic than the Fisher

projections, but you should be able to draw both structures. In general, if a substituents points to the right in the Fisher structure, it points down in the Haworth. if it points left, it points up.

In general, the OH on the a-anomer points down (ants down) while on the b-anomer it points up (butterflys up).

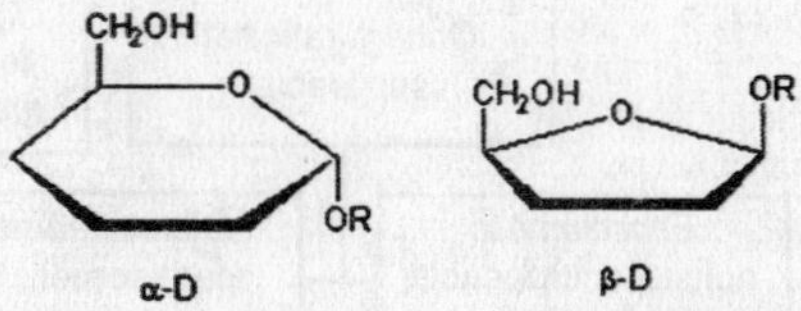

Fig. A more Rigorous view of the Relationship between the Anomeric OH and the OH on the Last Chiral C of a Sugar

The most common monosaccharides (other than glyceraldehyde and dihydroxyacetone) which you need to know are shown below.

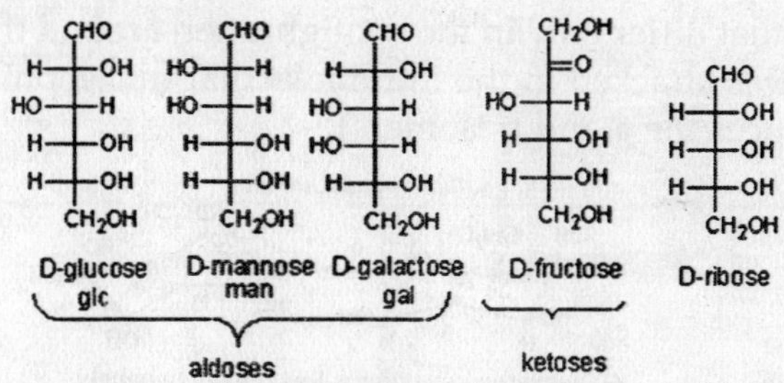

The mirror image of D-Glc is L-Glc. For common sugars, the prefix D and L refer to the cenre of asymmetry most remote from the aldehyde or ketone.

By convention, all chiral centers are related to D-glyceraldehyde, so sugar isomers related to D-glyceraldehyde at their last asymmetric cenre are D sugars.

ISOMERS

Sugars can exists as either configurational isomers (interconverted only by breaking covalent bonds) and conformational isomers.

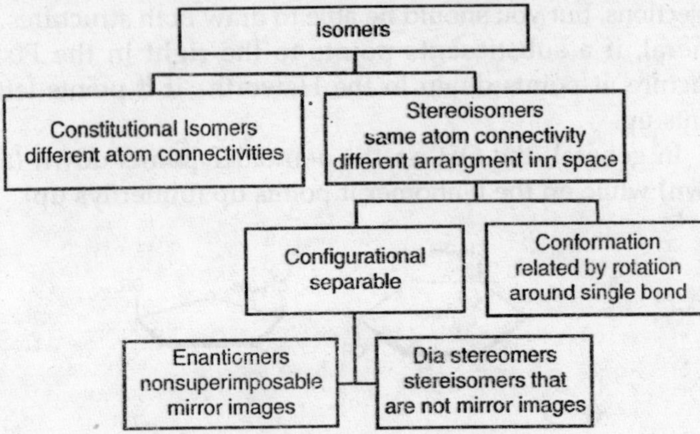

Fig. Different Types of Isomers

The configurational isomers include enantiomers (stereoisomers that are mirror images of each other), diastereomers (stereoisomers that are not mirror images), epimers (diastereomers that differ at one stereocenter) and anomers (a special form of stereoisomer, diastereomer and epimer that differ only in the configuration around the carbon which was attacked in the intramolecular nucleophilic attack to produces the a and b isomers).

Conformational Isomers

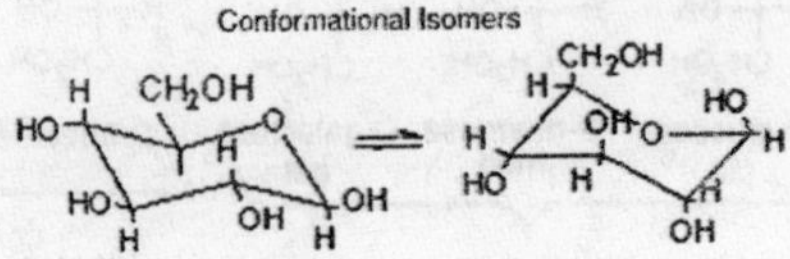

β-D-glucopyranose: chairnforms, boat form also is conformer

Monosaccharide Derivatives

Many derivatives of monosaccharides are found in nature. *These include*:

- Oxidized forms in which the aldehyde and/or alcohol functional groups are oxidized to carboxylic acids
- Phosporylated forms in which phosphate is added by ATP to form phosphoester derivatives
- Amine derivatives such as glucosamine or galactosamine
- Acetylated amine derivatives such as N-Acetyl-

GlcNAc (GlcNAc) or GalNAc

- Lactone forms (intramolecular esters) in which an OH group attacks a carbony C that was previously oxidized to a carboxylic acid
- Condensation products of sugar derivatives with lactate ($CH_3CHOHCO_2^-$) and pyruvate, ($CH_3COCO_2^-$), both from the glycolytic pathway, to form muramic acid and neuraminic acids, (also called sialic acids), respectively.

N-acetylmuramic acid, found in bacterial cell walls, consists of GlcNAc in ether link at C3 with lactate, while N-acetylneuraminic acid results from a intramolecular cyclization of an condensation product of ManNac and pyruvate.

CHEMISTRY OF SUGARS-HEMIACETALS, ACETALS, DISACCHARIDES,

Monosaccharides that contain aldehydes can cyclize through intramolecular nucleophilic attack of an OH at the carbonyl carbon in an addition reaction to form a hemiacetal (hemiketal if attack on a ketone).

On the addition of acid (which protonates the anomeric OH, forming water as a potential leaving group), another alcohol can add forming an acetal (or ketal from a ketone) with water leaving.

Aldehydes, Hemiacetals, and Acetals

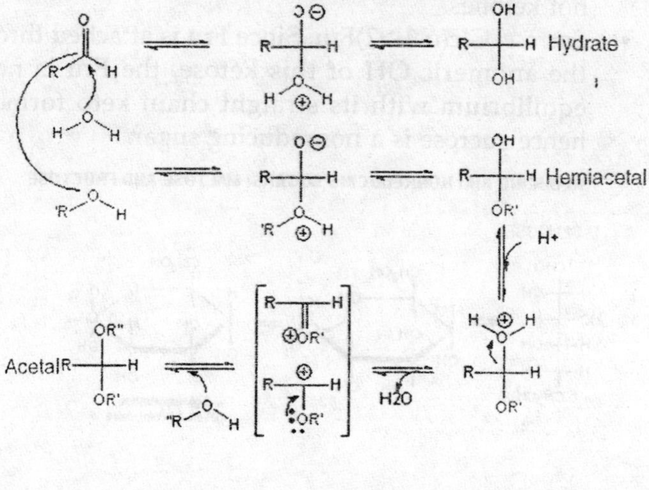

If the other alcohol is a second monosaccharide, a dissacharide results. The acetal (or ketal) link bonding to the two monosaccharides is called a glycosidic link. Links between the two sugars can be either a (if the OH on C1 involved in the glycosidic link is pointing down) or b (if the O on C1 involved in the glycosidic link is pointing up).

Since sugars contain so many OH groups which can act as the "secondΔ alcohol in acetal (or ketal) formation, links between sugars can be quite diverse. These include a and b forms of 1-2, 1-3, 1-4, 1-5, 1-6, 2-2, etc. links.

For example:

- *Lactose:* Gal(β 1->4)Glc Since Glc is attached to Gal through the OH on C4, its anomeric carbon, C1, could revert to the noncyclic aldehyde form. This aldehyde is susceptible to oxidation by reagents (Benedicts Solution - with citrate, Fehlings reagent - with tartrate) which are subsequently reduced. In both reagents, reducing sugars reduce a basic blue solution of $CuSO_4$ (Cu^{2+}) to form a brick red precipate of Cu_2O (Cu^{+}). Sugars (monosaccharides, dissacharides and polysaccharides which can form an aldehyde at C1 or have an a-hydroxymethyl ketone group which can isomerize to an aldehyde under basic conditions (such as fructose) are called reducing sugars. These oxidizing agent are mild and react with aldehydes and not ketones.
- *Sucrose:* Glc(α 1->2)Fru. Since Fru is attached through the anomeric OH of this ketose, the Fru is not in equilibrium with its straight chain keto form and hence sucrose is a nonreducing sugar.

REDUCING AND NONREDUCING SUGARS: MALTOSE AND FRUCTOSE

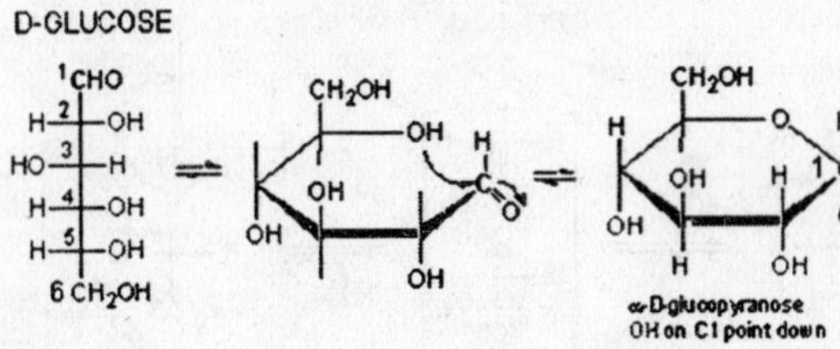

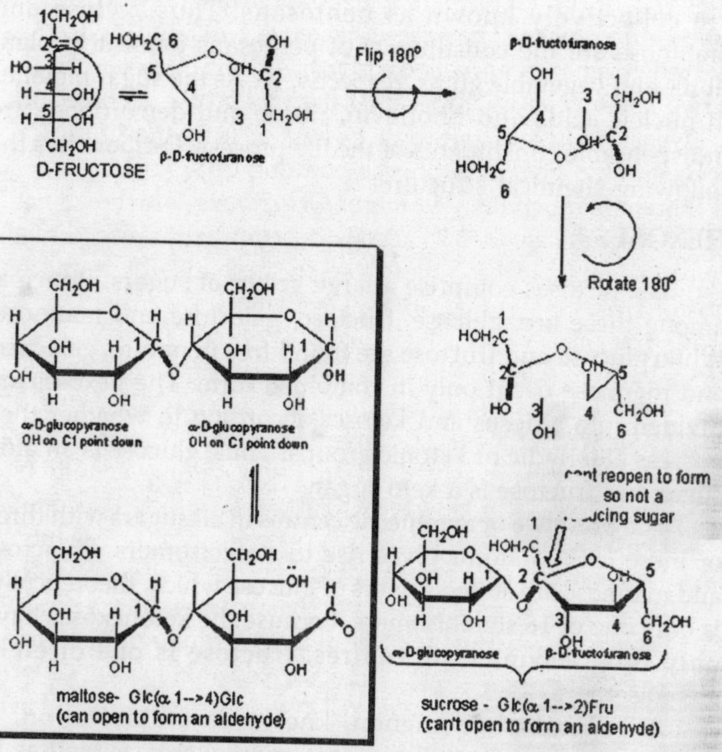

The glycosidic (acetal or ketal) link can be cleaved by hydrolysis, just as the peptide bond in proteins.

CLASSIFICATION OF CARBOHYDRATE

Carbohydrates are classified generally according to their degree of complexity. Hence, the free sugars such as glucose and fructose are termed monosaccharides; sucrose and maltose, disaccharides; and the starches and celluloses, polysaccharides. Carbohydrates of short chain lengths such as raffinose, stachyose and verbascose, which are three, four and five sugar polymers respectively, are classified as oligosaccharides.

PENTOSES

Pentoses are five-carbon sugars seldom found in the free state in nature. In plants they occur in polymeric forms and are collectively known as pentosans. Thus, xylose and arabinose are the constituents of pentosans present in plant fibres and vegetable gums, respectively. As the sugar moieties in nucleic acids and riboflavin, ribose and deoxyribose are indispensable constituents of the life process. D-ribose has the following chemical structure.

HEXOSES

The hexoses comprise a large group of sugars. Principal among these are: glucose, fructose, galactose and mannose. While glucose and fructose are found free in nature, galactose and mannose occur only in combined form. The hexoses are divided into aldoses and ketoses according to whether they possess aldehydic or ketonic groups. Thus, glucose is an aldo sugar and fructose is a keto sugar.

The presence of aymmetric centres in all sugars with three or more carbon atoms gives rise to stereoisomers. Galactose and mannose are stereoisomers of glucose which, theoretically, is only one of 16 stereoisomers. Because the ketohexoses have only three asymmetric centres, fructose is one of eight stereoisomers.

A general phenomenon, known as mutarotation, is observed in a variety of pentoses and hexoses as well as in certain disaccharides. For example, it has been established that two isomers of D-glucose exist, hence requiring an additional asymmetric centre in this sugar. It became apparent that D-glucose and most other sugars have cyclic structures.

The position of the hydroxyl group in relation to the ring oxygen characterizes this additional configurations modification. By convention, the positioning of the hydroxyl group on carbon atom 1 on the same side of the structure as the oxygen ring -modification; and, the positioning of the same hydroxyl group on the opposite side of the ring oxygen b-modification.

Carbohydrases, which catalyse the hydrolysis of glycosidic linkages of simple glycosides, oligosaccharides and polysaccharides often exhibit specificity with regard to substrate configuration. The specificity for enzyme hydrolysis of certain oligosaccharides helps to explain the poor utilization of this class of carbohydrates in fish nutrition.

Sugars containing the aldo or the keto group are capable of reducing copper in alkaline solutions to produce the brick-red colouration of cuprous ions. These sugars are called reducing sugars and the reaction, although not specific for reducing sugars, has use for both qualitative and quantitative determinations.

Glucose is widely distributed in small amounts in fruits, plant juices and honey. It is commercially produced by the acid or enzyme hydrolysis of grain and root starches. Glucose is of special interest in nutrition because it is the end-product of carbohydrate digestion in all non-ruminant animals including fish. Fructose is the only important ketohexose and is found in the free state alongside glucose in ripening fruits and honey. Combined with glucose it forms sucrose. Fructose is somewhat sweeter than sucrose and is produced in increasing quantities commercially as a sweetener.

Galactose occurs in milk in combination with glucose. It is also present in oligo-saccharides of plant origin, in combination with both glucose and fructose.

Mannose is present in some plant polysaccharides collectively termed mannans.

DISACCHARIDES

Disaccharides are condensation products of two molecules of monosaccharides. Sucrose is the predominant disaccharide occurring in the free form and is the principal substance of sugar cane and sugar beet. It is also formed during germination of legume seeds. Other common disaccharides are maltose and lactose.

Maltose is a dimer of glucose and lactose is a copolymer of galactose and glucose. The two molecules of glucose in maltose are held together in an a -1,4 glycosidic linkage whereas the two hexose entities of galactose are linked at the b -1,4 position. Glucose and fructose are combined in an a -1,2 linkage in sucrose. The abbreviated name of sucrose is D-Glu-(a, 1→ 2)-D-Fru.

OLIGOSACCHARIDES

The Oligosaccharides raffinose, stachyose and verbascose are present in significant quantities in legume seeds. Raffinose, which is the most widespread among the three, consists of one molecule of glucose linked to a molecule of sucrose at the α-1, 6 position. Its abbreviated chemical name is α-D-Gal (1→6) -α - D -Glu - (1 → 2) - β-D-Fru. Further chain elongation at the galactose end with another galactose molecule will yield stachyose.

These galactose-galactose linkages are all at the α-l,6 position and digestion of these Oligosaccharides by animals requires a highly specific enzyme not elaborated by the animals themselves but by certain bacteria present in the animals guts. The gradual disappearance of oligosaccharides from the cotelydons of legume seeds during germination is part of an intricate process beginning with uptake of water by the seed. This uptake of moisture releases gibberellic acid which in turn activates the DNA in the seed, thereby triggering the life cycle of the plant.

The DNA directs the production of α-galactosidase which is required for the hydrolysis of these Oligosaccharides. Any interference of the DNA transcription process blocks enzyme production and will be evidenced by continued senescence of the seed and persistence of oligosaccharides in the seed cotelydons.

POLYSACCHARIDES

The polysaccharides represent a large group of complex carbohydrates which are condensation products of undetermined numbers of sugar molecules. The various subgroups are rather ill-defined and there is a lack of agreement on their classification. Most polysaccharides are insoluble in water. Upon hydrolysis with acids or enzymes they eventually yield their constituent monosaccharides.

Starch is a high molecular weight polymer of D-glucose and is the principal reserve carbohydrate in plants. Most starches consist of a mixture of two types of polymers, namely; amylose and amylopectin. The proportion of amylose and amylopectin is generally one part of amylose and three parts of amylopectin. Enzymes capable of catalyzing the hydrolysis of starch are present in the digestive secretions of animals and fish within their cells.

The a-amylases which are found virtually in all living cells cleave the α-D-(1→4) linkages at random and bring about an eventual total conversion of the starch molecule into the reducing sugars. The principal α-amylases of animal origin are those produced in the salivary gland and the pancreas. Starch is insoluble in water and is stained blue by iodine.

Glycogen is the only complex carbohydrate of animal origin. It exists in limited quantities in liver and muscle tissues and acts as a readily available energy source. Dextrins are intermediate compounds resulting from incomplete hydrolysis or digestion of starch. The presence of α-D-(1→6) linkages in amylopectin and the inability of α-amylase to cleave these bonds give rise to low molecular weight carbohydrate segments called limit dextrins. These residues are acted upon primarily by acidophilic bacteria in the digestive tract.

Cellulose is made up of long chains of glucose units held together by β-D-(1→4) linkages. The enzymes which cleave these linkages are not ordinarily present in the digestive secretions of animals and fish although some species of shellfish are believed to elaborate cellulase, the enzyme which catalyzes the hydrolysis of cellulose. Cellulase producing micro-organisms present in the gut of herbivorous animals and fish impart to their host animals the ability to utilize as food the otherwise indigestible cellulose.

Other complex polysaccharides in common occurrence are the hemicelluloses and pentosans. Hemicellulose represents a group of carbohydrates including araban, xylan, certain hexosans and polyuronides. These substances are generally less resistant to chemical treatment and undergo some degree of enzymatic hydrolysis during normal digestive processes. Pentosans are polymers of either xylose or arabinose as constituents of plant structural material and vegetable gums, respectively. Arbohydrates are classified generally according to their degree of complexity. Hence, the free sugars such as glucose and fructose are termed monosaccharides; sucrose and maltose, disaccharides; and the starches and celluloses, polysaccharides. Carbohydrates of short chain lengths such as raffinose, stachyose and verbascose, which are three, four and five sugar polymers respectively, are classified as oligosaccharides.

GLYCOPROTEINS

For many years glycoproteins have been a subject of interest. However, it is in the second half of this century that they have aroused the interest of biochemists and biologists from a wide range of fields. This increased interest is partly due to the fact that glycoproteins were discovered to be abundant in living organisms. It is also due to the diverse functions of glycoproteins; glycoproteins appear in nearly every biological process studied. Many glycoproteins have structural functions. One of many instances is their role as a constituent of the cell wall. Glycoproteins also form connective tissues such as collagen. They are also found in gastrointestinal mucus secretions. Glycoproteins are used as protective agents and lubricants. They are also found abundantly in the blood plasma where they serve many functions.

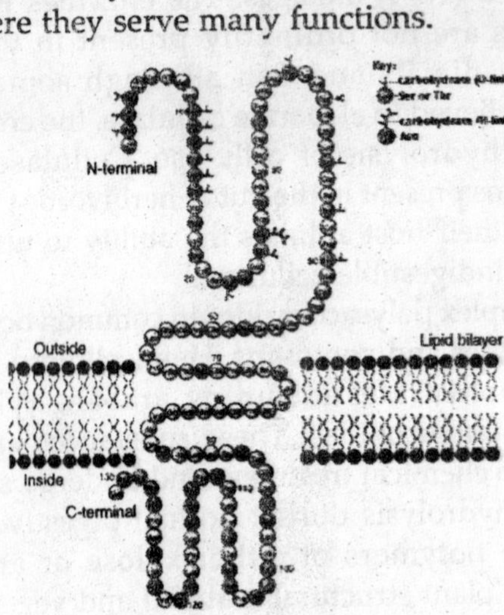

Fig. Structure of Glycoproteins

The diverse function of glycoproteins is a direct result of their structure. These macromolecules are composed of a peptide chain with one or more carbohydrate moieties. There are two broad categories of glycoprotein structure. The carbohydrates are either linked N-glycosidically or O-glycosidically to their constituent protein.

Within these broader categories, there can be fine structural differences which account for the large diversity of functions among glycoproteins. Controlling of glycoproteins is achieved through synthesis and degradation. Those processes are controlled by very specific enzymes. Not so much is yet researched on glycoprotein regulation in general.

STRUCTURE

Structurally, glycoproteins consist of a polypeptide covalently bonded to a carbohydrate moiety. The carbohydrate can make up anywhere from less than one per cent to more than 80 per cent of the total protein mass. The saccharide chains, referred to as glycans, can be linked to the polypeptide in two major ways.

The first class of glycoproteins are the O-linked glycans. These usually contain an N-acetylgalactosamine which is attached through a glycosidic bond to the O-terminus of either threonine or serine. The other class of glycoproteins are the N-linked glycans. These involve a glycosidic bond between N-acetylglucosamine and the N-terminus of an asparagine residue.

O-linked glycans consist of N-acetylgalactosamine attached to the O-terminus of a threonine or serine residue. N-acetylgalactosamine is simply a galactose molecule with an amine group covalently bonded to the second carbon. This amine group is bonded to a carboxyl group. N-acetylgalactosamine attaches to the carboxyl group of the amino acid through the hydroxyl group of its anomeric carbon.

Another type of O-linked glycan consists of a galactose or a glucosyl-galactose disaccharide linked to the hydroxyl of hydroxylysine. Yet another type of O-linkage involves the binding of arabinose to the hydroxyl of hydroxyproline. In all of the 0-linked glycans, there can be a variety of different monosaccharide or polysaccharide chains attached to the sugar that is bonded to the amino acid.

The other class of glycoproteins are N-linked glycans. These molecules consist of an N-acetylglucosamine bonded to the amide nitrogen of an asparagine molecule. N-acetylglucosamine is simply a glucose molecule which is bonded to an amine group. This amine group, in turn, is bonded to a hydroxyl group.

The N-acetylglucosamine is bonded to the asparagine through its anomeric carbon. The asparagine must be surrounded by a specific amino acid sequence, or sequon. This sequence is -X-Asn-X-Thr; the X can be any amino acid. A large variety of polysaccharide side chains can be linked to the N-acetylglucosamine. A typical polysaccharide chain is Man2 a(1-6)-Man B(1-4)-GlcNAc B(1-4)-GlcNAc B(1-N) Asn. Adding on to this structure can create many different N-linked glycans.

The carbohydrate chains of glycoproteins can play a role in the structure of the polypeptide. For example, in human immunoglobulins, the carbohydrate chain wraps around one of the protein domains. By doing so it prevents contact of this domain with the neighboring domain. An experiment that was done by Koide et al illustrates how the carbohydrate can change the overall structure of the polypeptide.

The carbohydrate side chains of a rabbit antibody were removed through glycosidase digestion. The result was that the domain where the carbohydrate had previously been attached could no longer perform its ordinary function. Because an immunoglobulin's function is determined to a large extent by its structure, it can be concluded that removing the carbohydrate affected the structure of the molecule.

FUNCTION

Because carbohydrates and proteins by themselves serve in a vast number of biological functions, it should not be surprising that linking the two together results in a macromolecule with an extremely large number of functions. Because of this and their biologically ubiquitous nature, the best way to go about exploring glycoprotein function is to break it down into categories that are fairly general. The following is an attempt to do this.

Structural: Glycoproteins are found throughout matrices. They act as receptors on cell surfaces that bring other cells and proteins together giving strength and support to a matrix

Proteoglycan: Linking glycoproteins cross links proteoglycan molecules and is involved in the formation of the ordered structure within cartilage tissue. In nerve tissue glycoproteins are abundant in gray matter and appear to be associated with synaptosomes, axons and microsomes. Prothrombin, thrombin and fibrinogen are all glycoproteins that play an intricate role in the blood clotting mechanism. In certain bacteria the slime layer that surrounds the outermost components of cell walls are made up of glycoproteins of high molecular weight. In addition to forming these s-layers, glycoproteins also function as bacterial flagella.

These are made up of bundles of glycoproteins protruding from the cell's surface. Their rotation provides propulsion. In plants, glycoproteins have roles in cell wall formation, tissue differentiation, embryogenesis and sexual adhesion

Protection: High molecular weight polymers called mucins are found on internal epithelial surfaces. They form a highly viscous gel that protects epithelium form chemical, physical and microbial disturbances. Examples of mucin sites are the human digestive tract, urinary tract and respiratory tracts. "Cervical mucinΔ is a glycoprotein found in the cervix of animals that regulates access of spermatozoa to the upper reproductive tract.

Recently it was discovered that mucins may be responsible for aiding in metastasis of transformed cancer cells Mucins are also found on the outer body surfaces of fish to protect the skin Not only does mucin serve the function of protection, but it also acts as a lubricant. Human lacrimal glands produce a glycoprotein which protects the corneal epithelium from desiccation and foreign particles. Human sweat glands secrete glycoproteins which protect the skin from the other excretory products that could harm the skin

Reproduction: Glycoproteins found on the surface of spermatozoa appear to increase a sperm cell's attraction for the egg by altering the electrophoretic mobility of the plasma membrane. Actual binding of the sperm cell to the egg is mediated by) -linked glycoproteins serving as receptors on the surface of each the two membranes.

The zona pellucida is an envelope made of glycoprotein that surrounds the egg and prevents polyspermy from occurring after the first sperm cell has penetrated the egg's plasma membrane Hen ovalbumin is a glycoprotein found in egg white that serves as a food storage unit for the embryo

Adhesion: Glycoproteins serve to adhere cells to cells and cells to substratum. Cell-cell adhesion is the basis for the development of functional tissues in the body. The interactions between cells is mediated by the glycoproteins on those cell's surfaces. In different domains of the body, different glycoproteins act to unite cells. For example, nerve cells recognize and bind to one another via the glycoprotein N-CAM (nerve cell adhesion molecule). N-CAM is also found on muscle cells indicating a role in the formation of myoneural junctions.

With cell-substratum adhesion, glycoproteins serve as cell surface receptors for certain adhesion ligands that mediate and coordinate the interaction of cells. Substrates with the appropriate receptor will bind to the cell related to that receptor. For example, a substrate containing the glycoprotein fibronectin will be recognized and adhered to by fibroblasts. The fibroblasts will then secrete adhesion molecules and continue to spread, producing a pericellular matrix.

Hormones: There are many glycoproteins that function as hormones such as human chorionic gonadotropin (HCG) which is present in human pregnancy urine. Another example is erythropoietin which regulates erythrocyte production

Enzymes: Glycoprotein enzymes are of three types. These are oxidoreductases, transferases and hydrolases

Carriers: Glycoproteins can bind to certain molecules and serve as vehicles of transport. They can bind to vitamins, hormones, cations and other substances.

Inhibitors: Many glycoproteins in blood plasma have shown antiproteolytic activity. For example, the glycoprotein a1-antichymotrypsin inhibits chymotrypsin.

Defence: In beetles pygidial glands secrete a glycoprotein disinfecting paste that covers the body and hardens. This shell provides protection against attack by bacteria and fungi

Freezing-point depression: Glycoproteins were found in the sera of antarctic fishes to decrease the freezing point due to their apparent interaction with water

Vision: In bovine visual pigment a glycoprotein forms the outer membranes of retinal rods

Immunological: The interaction of blood group substances with antibodies is determined by the glycoproteins on erythrocytes. Adding or removing just one monosaccharide from a blood group structure, the antigenicity and therefore a person's blood type can be altered Many immunoglobulins are actually glycoproteins Soluble immune mediators such as helper, suppressor and activator cell have been shown to bind to glycoproteins found on the surface of their target cells. B and T cells contain surface glycoproteins that attract bacteria to these sites and bind them. In much the same manner, glycoproteins can direct phagocytosis. Because the HIV virus recognizes the receptor protein CD4, it binds to helper T cells which contain it.

REGULATION/CONTROL

Regulation and control of glycoproteins is not as straight forward as some might think. "If someone was to understand regulation in glycoproteins he would have to look at the enzymes that are involved in the biosynthesis pathway of these molecules; what is affecting the enzyme activity (could be hormonal). The big interest in this area is in the cloning of the genes, the cloning of the enzymes that are involved in the synthesis of the oligosaccharide on the glycoproteins.

SYNTHESIS

The protein part of the glycoprotein is formed at the ribosomes, where all proteins are synthesized, on template represented by RNA and DNA. As a result, its structure can change only through the mutation of the genetic material of the cell. On the other hand, the carbohydrate component of a glycoprotein is not a product of the ribosome; it is synthesized somewhere in the cytoplasm, but the exact site of synthesis has not yet been established. Since it is not directly genetically controlled, the oligosaccharide part shows a much greater variation

In contrast with O-linked glycoproteins, where oligosaccharide assemble occurs on the polypeptide chain, N-linked glycoproteins assemble their oligosaccharide portions on a lipid linked intermediate, dolichol phosphate. The first step in oligosaccharide synthesis is the formation of that intermediate. In subsequent steps, sugars, the first being always N-acetylglucosamine are chain-like connected to dolichol phosphate.

One more GlcNAc and three more mannose sugars are linked to the first GlcNAc to form the typical core of N-linked glycoproteins. Addition of any other sugar can be in any possible combination, according to the desired resulting functions. The specificity of the enzymes is very important in the synthesis process. Every sugar added, is catalyzed by a different enzyme.

These group of enzymes are called glycosyltransferases The first addition of GlcNAc to dolichol phosphate is catalyzed by the specific enzyme example) that cleave peptide bonds and glycosidases enzymes that remove sugars one at a time from the end of an oligosaccharide chain. Both of these groups are contained in lysozomes. The lysosome attaches to a phagocyte, which has engulfed a substance that needs to be broken down and releases its enzymes in it Next, these enzymes begin their catalytic action. In degradation, ad in synthesis, the enzymes involved are very specific.

After the glycoprotein is broken down, its amino acid and sugar components are either metabolized or can be used in the formation of another glycoprotein. Enzymatic degradation can provide much information about the structure of the oligosaccharide chains, as well as about the carbohydrate peptide linkage. For example, if a glycoprotein is treated with mannosidase (removes mannose) and mannose is released, one can conclude that mannose residues were located at the periphery of the molecule since glycosidases remove sugars from the end of the oligosaccharide chain

Regulation and control is the organism's ultimate tool to monitor and adjust the production or degradation of different molecules. Like a factory, it would be useless and inefficient to produce excess products when there is no need for them. On the other hand, shortage of products could be a problem too. Hopefully, further research will put some light in the area of glycoprotein regulation.

GLYCOPROTEIN STRUCTURE AND STEROID HORMONE FUNCTION

Glycoproteins play an important part in hormone function. The action of hormones depends on the initial binding of the hormone to a protein receptor molecule. In many cases this molecule is a glycoprotein. Many hormones bind to receptors in the cell membrane; these hormones never actually enter the cell. Steroids, on the other hand, bind to an intracellular protein receptor. There is still controversy over whether the steroid hormone receptor is found in the nucleus or the cytosol. However, it is clear that after the steroid binds, the hormone-receptor complex moves to the nucleus.

The hormone binding domain of the receptor is found at the C-terminus. The amino acid sequence of this region is highly diverse; it is not conserved from one protein to the next. Binding of the hormone stimulates a conformational change in the hormone receptor. This change allows the hormone-receptor complex to bind to the DNA.

The DNA binding domain is highly conserved and is found within the central core of the protein. The central core contains a very basic amino acid sequence. In the cellular environment, these bases will tend to pick up a hydrogen and become positively charged. The positive charge will attract the hormone-receptor complex to the negatively charged DNA. Binding of the complex to the DNA stimulates transcription

SIMILARITIES BETWEEN GLYCOPROTEINS, CYTOCHROMES AND METALLOENZYMES

Cytochromes have features that are quite similar to transferrins, a particular class of glycoproteins. First of all, cytochromes make up some of the integral proteins found in membranes, as do glycoproteins. Second of all, the main function of cytochromes is the binding of ion cations. Via oxidoreduction with these cations of 2+state, cytochromes bind iron to a porphyrin prosthetic group, forming a heme group The function is to transport cations in and out of the cell. Transferrin is an iron binding protein as well.

It interacts with iron cations of the 2+state which also bind to a porphyrin group, Protoporphyrin IX, to form a heme group. Transferrins carry ions to many different locations in the body including in and out of cells Another group of molecules that serve a function related to glycoproteins are the metalloenzymes.

One class of glycoprotein enzymes, the oxidoreductases, contain an enzyme called chloroperoxidase. Like transferrin, chloroperoxidase contains protophyrin IX and forms the heme group with iron 3+ Eicosanoids, are also related to glycoproteins because they recognize the receptors on cells which are glycoproteins.

INHIBITION OF ASPARAGINE LINKED GLYCOPROTEINS BY ANTIBIOTICS

Synthesis of N-linked glycoproteins can be inhibited with the use of antibiotics (function of antibiotics). Examples are monensin, mevinolin and tunicamycin. Monensin, a monocovalent ionophore, inhibits the transport of proteins through the endoplasmic reticulum- Golgi complex. Mevinolin on the other hand, is an inhibitor of mevalonate synthesis. Mevalonate is the precursor of dolichol and other isoprenoid lipids Asparagine linked glycosylation involves the first step of addition of GlcNAc to dolichol diphosphate.

This reaction is catalyzed by the enzyme UDP-GlcNAc:dolichol phosphate N-acetylglucosamine-1-phosphate transferase or GPT. Tunicamycin is a specific inhibitor of GPT, therefore it can inhibit the synthesis of all N-linked glycoproteins. To give just an example of the effectiveness of tunicamycin, when it was injected to rats' hearts, it killed 12, 24, 50 and 60 h after the injection. Another connection could be seen with the function of the ribosomes, since the polypeptide chain of all glycoproteins.

PEPTIDOGLYCAN

Peptidoglycan, also known as murein, is a polymer consisting of sugars and amino acids that forms a mesh-like layer outside the plasma membrane of eubacteria. The sugar component consists of alternating residues of β-(1,4) linked N-acetylglucosamine and N-acetylmuramic acid residues. Attached to the N-acetylmuramic acid is a peptide chain of three to five amino acids.

The peptide chain can be cross-linked to the peptide chain of another strand forming the 3D mesh-like layer. Some Archaea have a similar layer of pseudopeptidog-lycan. Peptidoglycan serves a structural role in the bacterial cell wall, giving structural strength, as well as counteracting the osmotic pressure of the cytoplasm.

A common misconcep-tion is that peptidoglycan gives the cell its shape; however, whereas peptidoglycan helps maintain the structure of the cell, it is actually the MreB protein that facilitates cell shape. Peptidoglycan is also involved in binary fission during bacterial cell reproduction. The peptidoglycan layer is substantially thicker in Gram-positive bacteria (20 to 80 nm) than in Gram-negative bacteria (7 to 8 nm), with the attachment of the S-layer.

Peptidoglycan forms around 90% of the dry weight of Gram-positive bacteria but only 10% of Gram-negative strains. In Gram-positive strains, it is important in attachment roles and sterotyping purposes. For both Gram-positive and Gram-negative bacteria, particles of approximately 2 nm can pass through the peptidoglycan.

ANTIBIOTIC INHIBITION

Some antibacterial drugs such as penicillin interfere with

the production of peptidoglycan by binding to bacterial enzymes known as penicillin-binding proteins or transpeptidases. Penicillin-binding proteins form the bonds between oligopeptide crosslinks in peptidoglycan.

For a bacterial cell to reproduce through binary fission, more than a million peptidoglycan subunits (NAM-NAG+oligopeptide) must be attached to existing subunits. Mutations in transpeptidases that lead to reduced interactions with an antibiotic are a significant source of emerging antibiotic resistance.

Considered the human body's *own antibiotic*, lysozymes found in tears work by breaking the α-(1,4)-glycosidic bonds in peptidoglycan and thereby destroying many bacterial cells. Antibiotics such as penicillin commonly target bacterial cell wall formation (of which peptidoglycan is an important component) because animal cells do not have cell walls.

Structure

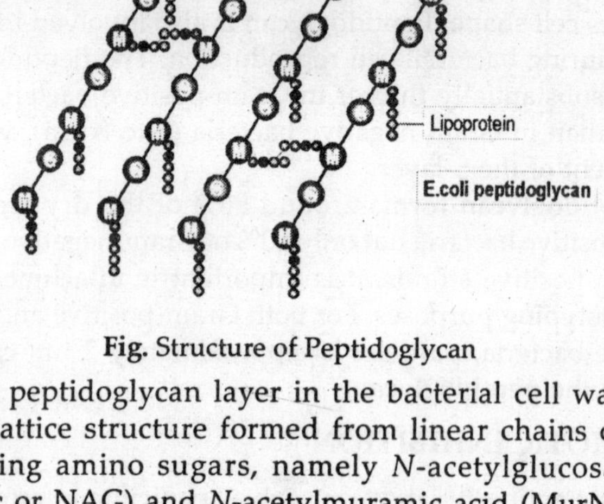

Fig. Structure of Peptidoglycan

The peptidoglycan layer in the bacterial cell wall is a crystal lattice structure formed from linear chains of two alternating amino sugars, namely *N*-acetylglucosamine (GlcNAc or NAG) and *N*-acetylmuramic acid (MurNAc or

NAM). The alternating sugars are connected by a α-(1,4)-glycosidic bond. Each MurNAc is attached to a short (4- to 5-residue) amino acid chain, normally containing D-alanine, D-glutamic acid and mesodiaminopimelic acid.

These three amino acids do not occur in proteins and are thought to help protect against attacks by most peptidases. Cross-linking between amino acids in different linear amino sugar chains by an enzyme known as transpeptidase result in a 3-dimensional structure that is strong and rigid. The specific amino acid sequence and molecular structure vary with the bacterial species.

LIPOPOLYSACCHARIDES

LIPOPOLYSACCHARIDES

These complex compounds (LPS) are the endotoxic O-antigens found in the cell walls of Gram-negative bacteria (S-lipopolysaccharides) but are also present in one fungus, they may cause several pathophysiological symptoms, such as fever, diarrhea, blood pressure decrease, septic shock and death.

A lipid part (Lipid A) forms a complex with a core polysaccharide part through a glycosidic linkage. The core part is linked to a third external region of a highly immunogenic and variable O-chain polysaccharide or O-antigen made up of repeating oligosaccharide units. The latter region of the LPS molecule is responsible for bacterial serological strain specificity.

Lipid A consists of a backbone of b-1,6-glucosaminyl-glucosamine with two phosphoester groups in the 1-position of glucosamine I and in the 4-position of glucosamine II. The 3-position of glucosamine II forms the acid-labile glycosidic linkage to the long-chain polysaccharide. The other groups are substituted with hydroxylated fatty acids as hydroxymyristate (two ester-linked and two amide-linked) and normal fatty acids. The fatty acid profiles of lipopolysaccharides differ markedly from those of the phospholipids from the same species.

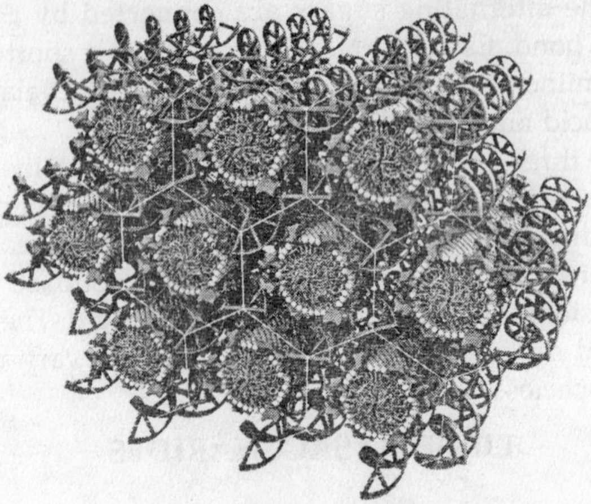

Fig. Structure of Lipid A of *Escherichia*

Although the hydroxylated fatty acids are generally 3-hydroxyalkanoic acids, 2-hydroxyalkanoic acids are also common. Exceptionally, some bacteria seem to lack hydroxy acids such as some species of *Gardnerella, Thermus, Thiobacillus, Bacteroides, Sorangium, Achromobacter* and *Brucella*. In most organisms, the major hydroxy acids are straight-chain, even-carbon compounds (from 12 to 18 carbon atoms).

Odd-carbon acids are minor components (13 to 17 carbon atoms) in some enterobacteria, some *Vibrio*, some Pseudomonads and *Bordetella*. They are also found in *Pectinatus, Veillonella* and *Selenomonas*. Branched-chain hydroxy fatty acids are also present in *Pseudomonas, Moraxella, Desulfovibrio, Capnocytophaga, Flavobacterium, Legionella* species (3-OH i-11:0 up to3-OH i-17:0).

Lipopolysaccharides are the major toxins of Gram-negative bacteria (endotoxin). The only Gram-positive bacteria that possesses LPS is *Listeria monocytogenes*, an infective agent in milk. While the polysaccharide is non-toxic, the lipid A part is responsible for the toxic activities of these bacteria which result in septic shock.

CARBOHYDRATE METABOLISM IN FISH

Much of the carbohydrates that enter the diets of animals,

including fish, is of plant origin. Carnivorous fish like the Atlantic salmon and the Japanese yellowtail, therefore, deal with little carbohydrate. Indeed, experiments have shown that these species are ill-equipped to handle significant quantities of raw carbohydrate, in their diets.

On the other hand, omnivores such as the common carp and the channel catfish are able to digest fair amounts of carbohydrates in their diets. The grass carp, a herbivore, subsists primarily on a vegetarian diet.

DIGESTION, ABSORPTION AND STORAGE

The ability of animals to assimilate starch depends on their ability to elaborate amylase. All species of fish have been shown to secrete a -amylase. It has also been demonstrated that activity of this enzyme was greatest in herbivores. In carnivores such as the rainbow trout and sea perch, amylase is primarily of pancreatic origin whereas in herbivores the enzyme is widespread throughout the entire digestive tract.

In Tilapia mossambica the pancreas has been shown to be the site of greatest amylase activity followed by the upper intestine. Although the digestion of starch and dextrin by the carnivorous rainbow trout was shown to decrease progressively as levels of the carbohydrates were increased beyond the 20 per cent level, the fish could effectively utilize up to 60 per cent glucose, sucrose or lactose in the diet. This demonstrates that, contrary to earlier belief, carnivorous fish are capable of efficiently utilizing simple carbohydrate as a primary energy source.

The crystalline structure of starch appears also to influence its attack by amylase as evidenced by the two-fold increase in metabolizable energy content of fully cooked (gelatinized) maize in feeding trials with channel catfish. Rainbow trout have also been shown to have a higher tolerance for carbohydrate (present as wheat starch) in the diet when it was cooked.

The process of gelatinization involves both heat and water. If an aqueous suspension of starch is heated, the granules do not change in appearance until a certain critical temperature is reached. At this point some of the starch granules swell and simultaneously lose their crystallinity. The critical temperature is that at which hydrogen bonds of the starch molecule loosen to permit complete hydration, leading to a phenomenon known as "swelling.

Alpha-amylase, promotes a more or less random fragmentation of the starch molecule by hydrolyzing at the α -D-(l→4) glucosidic bonds in the inner and outer chains of the compound. The result of complete hydrolysis of the amylose component are maltose and D-glucose, while the amylopectin component is reduced to maltose, D-glucose and branched limit dextrins. As a consequence of these action patterns by a -amylase on starch, other enzymes are needed for complete hydrolysis of starch to D-glucose in fish. In this regard, it has been demonstrated that even the carnivorous sea bream possess the ability to digest maltose.

On the other hand, cellulase and a -galactosidase have not been shown to be secreted by fish although cellulase of bacterial origin is present in the gut of most species of carps. The lack of a -galactosidase may partly explain the poor response by fish to dietary soybean meal which contains significant levels of the galactosidic oligosaccharides raffinose and stachyose. As has been pointed out earlier, these oligosaccharides do undergo enzymatic hydrolysis during the germination process to yield galactose and sucrose.

It would, therefore, appear that the nutritive value of soybean meal will be enhanced if the bulk of this indigestible starch is first transformed.

This can be achieved by soaking the beans for 48 hours prior to processing for meal production. It should also be pointed out that the nutritive value of pulses and other legume seeds can likewise be improved for fish since oligosaccharides constitute a large portion of the carbohydrates in legume seeds. Data on glucose absorption by fish are scanty. Work with goldfish has shown that active transport of glucose is coupled with Na^+ transport as in most mammals. It is generally believed that absorption takes place on the mucosal surface of intestinal cells.

The mono-saccharides which result from carbohydrate digestion consist primarily of glucose, fructose, galactose, mannose, xylose and arabinose. Although the rates of absorption of these sugars have been determined for many land mammals, similar information for fish is not available.

Glucose does not appear to be a superior energy source for fish over protein or fat although digestible carbohydrates do spare protein for tissue building. Also, unlike in mammals, glycogen is not a significant storage depot of energy despite evidence of an active and reversible Emden-Meyerhoff pathway in fish. The more efficient metabolism of amino acids over glucose for energy could be due to the ability of fish to excrete nitrogenous waste as ammonia from their gills without the high cost of energy in converting the waste to urea.

OTHER FACTORS AFFECTING METABOLISM

Apart from genetic adaptation, climatic factors also play an important role in carbohydrate metabolism in fish. Acclimation in fish, in essence, reflects enzyme acclimation, since the animal's ability to survive depends largely upon its ability to carry out normal metabolic functions. Some enzymes for metabolic acclimation show good compensation while others do not.

The enzymes associated with energy liberation (enzymes of glycolysis, pentose shunt, tricarboxylic acid cycle, electron transport and fatty acid oxidation) exhibit temperature compensation whereas, those enzymes dealing largely with the degradation of metabolic products show poor or reverse compensation

Table. Enzymes Subject to Metabolic Acclimation

Enzymes exhibiting compensation	Enzymes exhibiting reverse or no compensation
Phosphofructokinase	Catalase
Aldolase	Peroxidase
Lactic dehydrogenase	Acid phosphatase
6-phosphogluconate dehydrogenase	D-amino acid oxidase

Succinic dehydrogenase	Mg-ATP ase
Malic dehydrogenase	Choline acetyl transferase
Cytochrome oxidase	Acetylcholine esterase
Succinate-cytochrome C reductase	Alkaline phosphatase
NAD-cytochrome C reductase	Allantoinase
Aminoacyl transferase	Uricase
Na-K-ATPase	Amylase
	Lipase
protease	Malic enzyme
	Glucose-6-phosphate dehydrogenase

It is interesting to note that two key enzymes involved in carbohydrate metabolism, amylase and glucose-6-phosphate dehydrogenase, together with an enzyme involved in fat digestion, lipase, show no temperature compensation. It is not certain if this is in any way connected with the cessation of feeding by fish at low temperatures.

The molecular mechanism of thermal acclimation are not well understood and may consist of changes in synthesis or amounts of a given enzyme. Differences in kinetics, changes in the proportion of isoenzymes suitable for particular temperatures and changes in co-factors such as lipids, co-enzymes, or other factors such as pH and ions may be important in the animal's adjustment to temperature changes.

ENERGY TRANSFORMATION

Despite species differences in the tolerance of dietary carbohydrates it is generally believed that the principal end-product of carbohydrate digestion, glucose, is metabolized in a manner prevailing in all cells, i.e., via the reversible Emden-Meyerhoff pathway.

In this pathway, glucose has only one principal fate: phosphorylation to glucose-6-phosphate. The major metabolic transformations are depicted as follows: All transformations proceed with a loss of free energy. Thus, the formation of two moles of lactate from glucose-6-phosphate occurs with free energy change of D G^o = -22000 cal/mole. The net result is the formation of four molecules of ATP. A functional reversal of this transformation can only occur via a different sequence requiring the input of six ATP molecules per mole of glucose-6-phosphate recovered.

Cells do not store glucose or glucose-6-phosphate. The readily available storage form is glycogen which is made from glucose-1-phosphate by one pathway and returned by another. Although in mammalian cells glucose-6-phosphate is transformed into fatty acids, such transformation does not appear to take place in fish.

Studies with the common carp indicate that the precursor for lipogenesis is citrate formed when amino acids are actively metabolized through the tricarboxylic acid cycle. The major form of utilizable energy in all cells is ATP. In most cells this energy currency is generated by the oxidation of NADH by the mitochondrial electron-transport systems. The reductants of NAD^+ for this process are intermediates derived from the TCA cycle and fatty acids.

TCA Cycle

The TCA cycle showing enzymes, substrates and products. The abbreviated enzymes are: IDH = *isocitrate dehydrogenase* and α-KGDH = *α-ketoglutarate dehydrogenase*. The GTP generated during the *succinate thiokinase* (succinyl-CoA synthetase) reaction is equivalent to a mole of ATP by virtue of the presence of *nucleoside diphosphokinase*.

The 3 moles of NADH and 1 mole of $FADH_2$ generated during each round of the cycle feed into the oxidative phosphorylation pathway. Each mole of NADH leads to 3 moles of ATP and each mole of $FADH_2$ leads to 2 moles of ATP. Therefore, for each mole of pyruvate which enters the TCA cycle, 12 moles of ATP can be generated.

CITRATE SYNTHASE (CONDENSING ENZYME)

The first reaction of the cycle is condensation of the methyl carbon of acetyl-CoA with the keto carbon (C-2) of oxaloacetate (OAA) to form citrate The standard free energy of the reaction, -8.0 kcal/mol, drives it strongly in the forward direction. Since the formation of OAA from its precursor is thermodynamically unfavorable, the highly exergonic nature of the *citrate synthase* reaction is of central importance in keeping the entire cycle going in the forward direction, since it *drives* oxaloacetate formation by mass action principals.

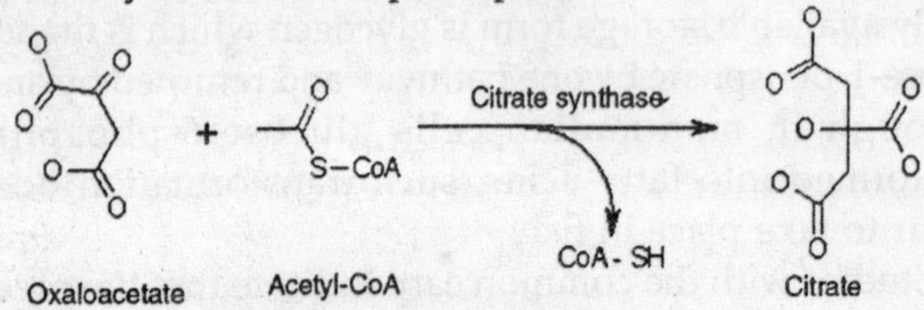

When the cellular energy charge increases the rate of flux through the TCA cycle will decline leading to a build-up of citrate. Excess citrate is used to transport acetyl-CoA carbons from the mitochondrion to the cytoplasm where they can be used for fatty acid and cholesterol biosynthesis. Additionally, the increased levels of citrate in the cytoplasm activate the key regulatory enzyme of fatty acid biosynthesis, acetyl-CoA carboxylase (ACC) and inhibit PFK-1. In non-hepatic tissues citrate is also required for ketone body synthesis.

ACONITASE

The isomerization of citrate to isocitrate by ***aconitase*** is stereospecific, with the migration of the -OH from the central carbon of citrate (formerly the keto carbon of OAA) being always to the adjacent carbon which is derived from the methylene (-CH_2-) of OAA.During the reaction the intermediate linked to the enzyme *cis*-aconitate is produced. The stereospecific nature of the isomerization determines that the CO_2 lost, as isocitrate is oxidized to succinyl-CoA, is derived from the oxaloacetate used in citrate synthesis.

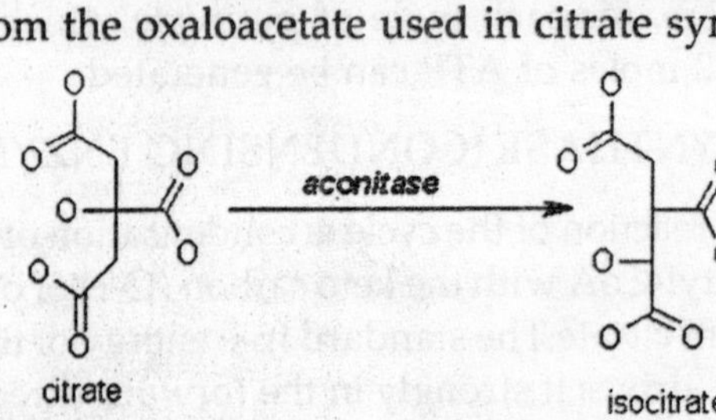

Aconitase is one of several mitochondrial enzymes known

as non-heme-iron proteins. These proteins contain inorganic iron and sulfur, known as iron sulfur centers, in a coordination complex with cysteine sulfurs of the protein.

There are two prominent classes of non-heme-iron complexes, those containing two equivalents each of inorganic iron and sulfur Fe_2S_2 and those containing 4 equivalents of each Fe_4S_4. Aconitase is a member of the Fe_4S_4 class. Its iron sulfur centers are often designated as $Fe_4S_4Cys_4$, indicating that 4 cystine sulfur atoms are involved in the complete structure of the complex. In iron sulfur compounds the iron is generally involved in oxidation-reduction events.

ISOCITRATE DEHYDROGENASE

Isocitrate is oxidatively decarboxylated to a-ketoglutarate by isocitrate dehydrogenase, (IDH).

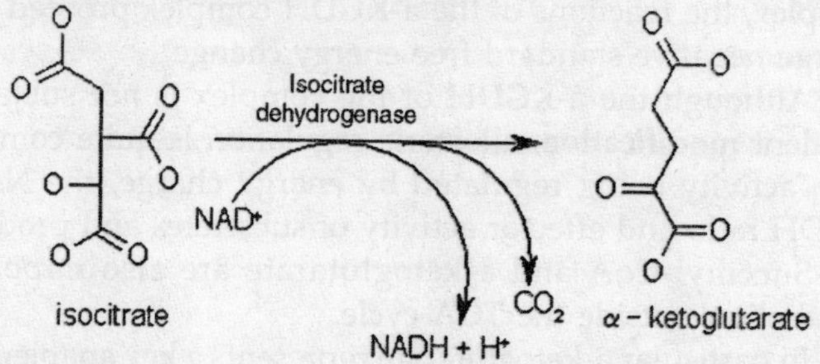

There are two different IDH enzymes. The IDH of the TCA cycle uses NAD^+ as a cofactor, whereas the other IDH uses $NADP^+$ as a cofactor. Unlike the NAD+-requiring enzyme, which is located only in the mitochondrial matrix, the $NADP^+$-requiring enzyme is found in both the mitochondrial matrix and the cytosol. IDH catalyzes the rate-limiting step, as well as the first NADH-yielding reaction of the TCA cycle. The CO_2 produced by the IDH reaction is the original C-1 of the oxaloacetate used in the citrate synthase reaction. It is generally considered that control of carbon flow through the cycle is regulated at IDH by the powerful negative allosteric effectors NADH and ATP and by the potent positive effectors; isocitrate, ADP and AMP. From the latter it is clear that cell energy charge is a key factor in regulating carbon flow through the TCA cycle.

α-KETOGLUTARATE DEHYDROGENASE

α-ketoglutarate is oxidatively decarboxylated to succinyl-CoA by the *a-ketoglutarate dehydrogenase* (α-KGDH) complex.

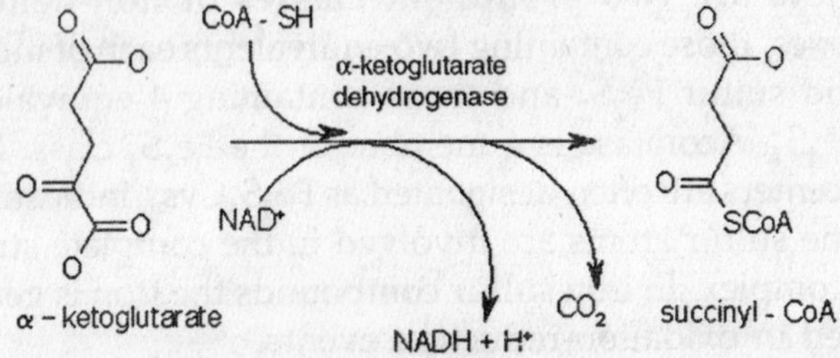

This reaction generates the second TCA cycle equivalent of CO_2 and NADH. This multienzyme complex is very similar to the PDH complex in the intricacy of its protein makeup, cofactors and its mechanism of action. Also, as with the PDH complex, the reactions of the a-KGDH complex proceed with a large negative standard free energy change.

Although the a-KGDH of the complex is not subject to covalent modification, allosteric regulation is quite complex, with activity being regulated by energy charge, the NAD^+/NADH ratio and effector activity of substrates and products.

Succinyl-CoA and a-ketoglutarate are also important metabolites outside the TCA cycle.

In particular,a-ketoglutarate represents a key anapleurotic metabolite linking the entry and exit of carbon atoms from the TCA cycle to pathways involved in amino acid metabolism. α-ketoglutarate is also important for driving the malate-aspartate shuttle. Succinyl-CoA, along with glycine, contributes all the carbon and nitrogen atoms required for the synthesis of protoporphyrin heme biosynthesis and for non-hepatic tissue utilization of ketone bodies.

SUCCINYL COA SYNTHETASE

The conversion of succinyl-CoA to succinate by succinyl CoA synthetase involves use of the high-energy thioester of succinyl-CoA to drive synthesis of a high-energy nucleotide phosphate, by a process known as substrate-level phosphorylation.

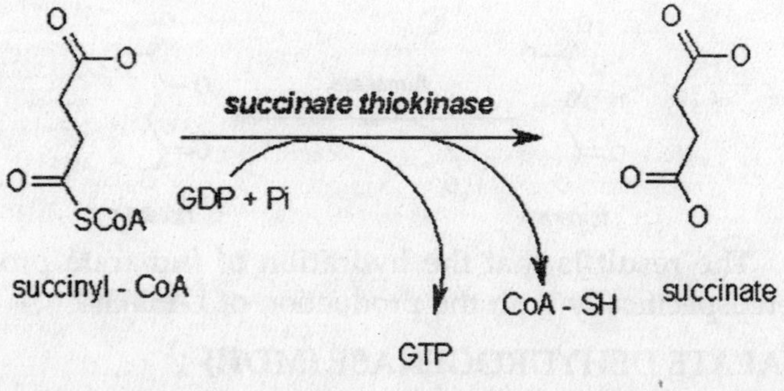

In this process a high energy succinyl-phosphate intermediate is formed, with the phosphate subsequently being transferred to GDP.

Mitochondrial GTP is used in a trans-phosphorylation reaction catalyzed by the mitochondrial enzyme nucleoside diphospho kinase to phosphorylate ADP, producing ATP and regenerating GDP for the continued operation of succinyl CoA synthetase.

SUCCINATE DEHYDROGENASE (SDH)

Succinate dehydrogenase catalyzes the oxidation of succinate to fumarate with the sequential reduction of enzyme-bound FAD and non-heme-iron.

In mammalian cells the final electron acceptor is coenzyme Q_{10} (CoQ_{10}), a mobile carrier of reducing equivalents that is restricted by its lipophilic nature to the lipid phase of the mitochondrial membrane.

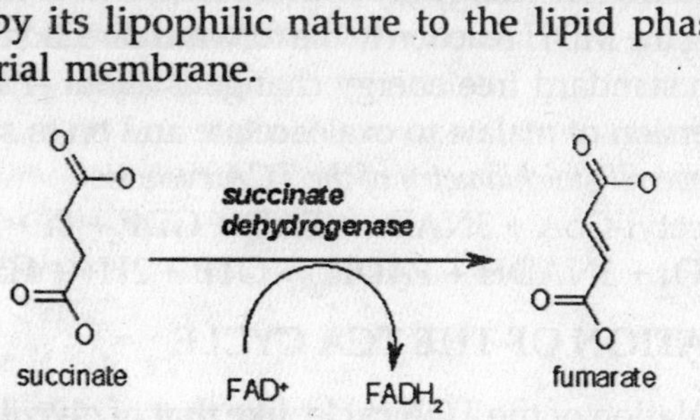

FUMARASE (FUMARATE HYDRATASE)

The fumarase-catalyzed reactions specific for the *trans* form of fumarate.

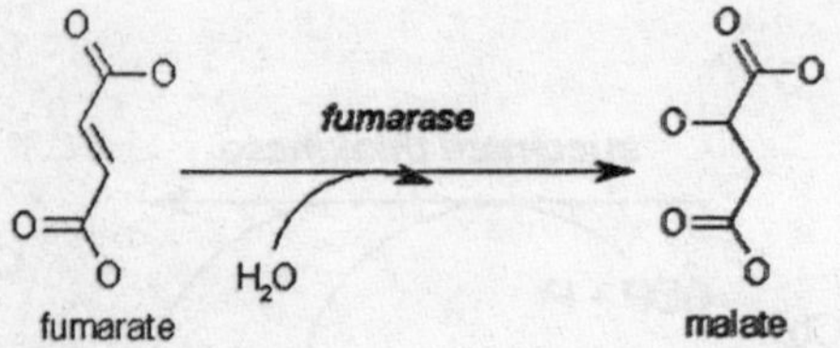

The result is that the hydration of fumarate proceeds stereospecifically with the production of L-malate.

MALATE DEHYDROGENASE (MDH)

L-malate is the specific substrate for MDH, the final enzyme of the TCA cycle. The forward reaction of the cycle, the oxidation of malate yields oxaloacetate(OAA).

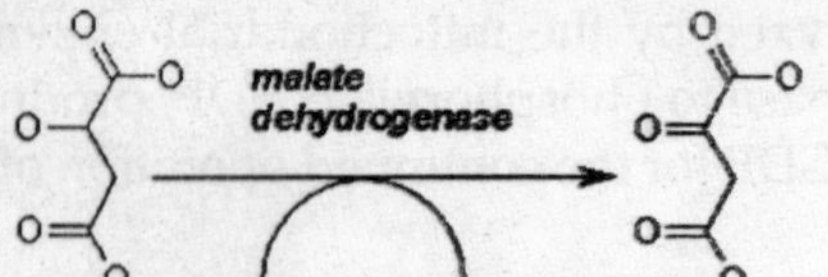

In the forward direction the reaction has a standard free energy of about +7 kcal/mol, indicating the very unfavorable nature of the forward direction. As noted earlier, the citrate synthase reaction that condenses oxaloacetate with acetyl-CoA has a standard free energy of about -8 kcal/mol and is responsible for pulling the MDH reaction in the forward direction. The overall change in standard free energy change is about -1 kcal/mol for the conversion of malate to oxaloacetate and on to succinate.

The overall stoichiometry of the TCA cycle is:

$$\text{acetyl-CoA} + 3NAD^+ + FAD + GDP + Pi + 2H_2O \rightarrow$$
$$2CO_2 + 3NADH + FADH_2 + GTP + 2H^+ + HSCoA$$

REGULATION OF THE TCA CYCLE

Regulation of the TCA cycle. like that of glycolysis, occurs at both the level of entry of substrates into the cycle as well as at the key reactions of the cycle. Fuel enters the TCA cycle primarily as acetyl-CoA. The generation of acetyl-CoA from carbohydrates is, therefore, a major control point of the cycle.

This is the reaction catalyzed by the PDH complex. By way of review, the PDH complex is inhibited by acetyl-CoA and NADH and activated by non-acetylated CoA (CoASH) and NAD^+. The pyruvate dehydrogenase activities of the PDH complex are regulated by their state of phosphorylation. This modification is carried out by a specific kinase (PDH kinase) and the phosphates are removed by a specific phosphatase (PDH phosphatase). The phosphorylation of PDH inhibits its activity and, therefore, leads to decreased oxidation of pyruvate. PDH kinase is activated by NADH and acetyl-CoA and inhibited by pyruvate, ADP, CoASH, Ca^{2+} and Mg^{2+}. The PDH phosphatase, in contrast, is activated by Mg^{2+} and Ca^{2+}.

Since three reactions of the TCA cycle as well as PDH utilize NAD^+ as co-factor it is not difficult to understand why the cellular ratio of NAD^+/NADH has a major impact on the flux of carbon through the TCA cycle.

Substrate availability can also regulate TCA flux. This occurs at the *citrate synthase* reaction as a result of reduced availability of oxaloacetate. Product inhibition also controls the TCA flux, e.g. citrate inhibits *citrate synthase,* a-KGDH is inhibited by NADH and succinyl-CoA. The key enzymes of the TCA cycle are also regulated allosterically by Ca^{2+}, ATP and ADP.

THE ELECTRON TRANSPORT CHAIN

$$2NADH + 2H^+ + O_2 \Longrightarrow 2NAD^+ + 2H_2O$$

The oxidation of NADH occurs in a stepwise manner involving an electron transport chain. Reducing equivalents go via a series of carriers to oxygen. Electrons flow downhill, energetically. Energy released in this process is used to generate the Proton Motive Force which is used by the cell to make ATP.

COMPONENTS OF THE ELECTRON TRANSPORT CHAIN (ETC)

The components of the ETC are membrane bound. Their redox potentials vary between NAD (Eo = -320mV) and O_2 (+820mV). Note that the redox potentials of different cytochromes and flavoproteins etc. may differ even though they have the same prosthetic groups.

The surrounding protein modifies the Eo of the heme and flavin groups. The mitochondria! ETC is fixed in composition and is longer and generates more energy than that of *E. coli*. In contrast, many bacteria, including *E. coli* have multiple alternative ETCs produced in response to different environmental conditions.

NADH Dehydrogenase

Found in the cytoplasmic membrane with other components of the ETC. *E. coli* contains two alternative NADH dehydrogenases: NADH-DH I contains one FAD, several FeS groups and one bound ubiquinone(UQ) per enzyme. This isoenzyme is coupled and generates PMF. Encoded by the gene *nuo* (NADH UQ oxidoreductase). Reducing equivalents go from NADH Ø FAD Ø FeS Ø UQ. NADH-DH II has FAD & UQ but contains no FeS clusters and is not coupled and cannot generate PMF. Encoded by *ndh*. It is used when there is surplus NADH to be oxidized but the cells have enough energy. The mitochondrial enzyme resembles NADH I except that it has FMN instead of FAD.

Ubiquinone

All ETCs contain quinones. Quinones are hydrogen carriers and carry 2H at a time. The structure of the quinone nucleus varies: ubiquinone (respiration) & plastoquinone (photosynthesis) each have a single aromatic ring whereas menaquinone (anaerobic respiration) has a double ring. Ubiquinone is sometimes called coenzyme Q and its side chain (R-group) is made of 6-10 isoprenoid (C5) units depending on the organism. In*E. coli* n = 8 and in mitochondria n = 10.

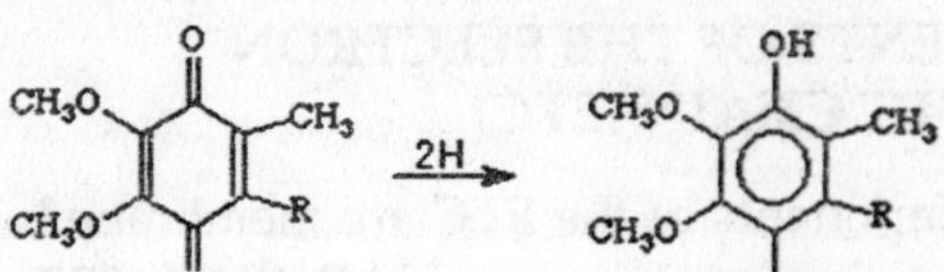

UBIQUINONE UBIQUINOL

There is more ubiquinone than all other ETCcomponents together. A small portion of the ubiquinone is tightly bound to ETC proteins and the rest is free to diffuse around in the membrane - the so-called ubiquinone pool.

The R-group = -[CH_2-CH=C(CH_3)-CH_2]n-H I

Iron Sulfur Proteins (FeS Proteins)

These have iron but no heme group, instead the iron is linked to acid-labile, i.e. inorganic, sulfur. Sometimes called non-heme-iron (NHI) proteins. The FeS groups are electron carriers. Each FeS group carries only one electron even though it may have 2 or 4 iron atoms. The electron is shared among the irons:

$$e + Fe^{2+} = Fe^{3+}$$

In *E. coli,* FeS groups are found in NADH-dehydrogenase-I and in the cytochrome b region of the chain. In mitochondria there are at least 7 FeS centers. Four are associated with mitochondrial NADH-dehydrogenase two are associated with cytochrome b and one with cytochrome c1.

Several variant structures, the most common are the flat Fe_2S_2 and the cube shaped Fe_4S_4 groups. Both are bound to the protein by four cysteine residues.

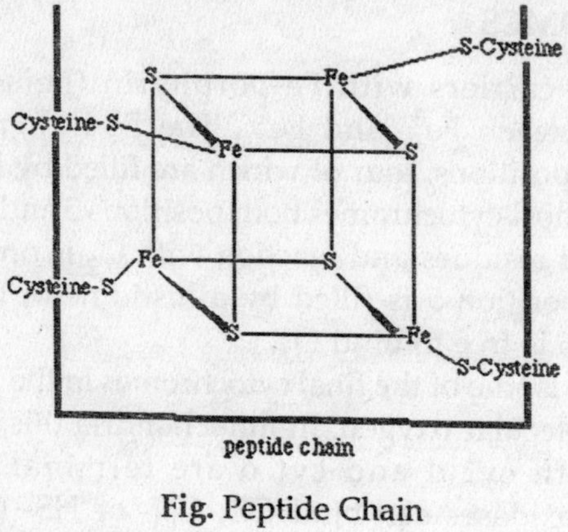

Fig. Peptide Chain

FLAVOPROTEINS

The NADH-dehydrogenases are flavoproteins. Several other flavoproteins join the ETC as side branches at ubiquinone, thus missing coupling site #1. Examples are succinate dehydrogenase(SDH), lactate dehydrogenase(LDH) and glycerol-P dehydrogenase. Note that *E. coli* contains two types of LDH:

- Soluble, NAD-linked, converts pyruvate to lactate in fermentation
- Membrane-bound, FAD-linked, oxidizes lactate to pyruvate in air and feeds into ETC. There are actually two FAD-linked isoenzymes, one each for the D- & L- isomers of lactate.

Flavoproteins may also contain FeS groups (e.g., SDH, NADH-DH I). Usually FAD or FMN is not bound covalently, but occasionally FAD may be bonded to a histidine (e.g. in SDH).

Flavoprotein dehydrogenases are linked to the ETC and are not reoxidised by molecular oxygen. In contrast, flavoprotein oxidases are not linked to the ETC and may be reoxidized directly by O_2 e.g., aldehyde oxidase, glucose oxidase.

Facultative anaerobes such as *E. coli* have few oxidase type flavoproteins.

CYTOCHROMES

Electron carriers with Fe-porphyrin (heme) groups. Alternate between Fe^{3+} and Fe^{2+}. The Fe in heme has six coordination positions, four of which are filled by N-atoms of the heme. In most cytochromes both positions 5 and 6 are filled by amino acid residues and reaction with O_2 is prevented. In hemoglobin position 5 is filled by a histidine of the protein and position 6 is free to bind O_2.

The same is true of the final cytochromes in the ETC which react with molecular oxygen. In mitochondria this is cyt a/a3, in *E. coli* both cyt d and cyt o are terminal oxidases. Cytochrome oxidases can bind CO, CN- or HS- (from H_2S) instead of O_2 and the enzyme activity is then killed.

In mitochondria and many obligately aerobic bacteria there is a full set of cytochromes: Electrons go from UQ to cyt b to cyt c1 to cyt c to cyt a/a3. Cyt a/a3 is known as cytochrome c oxidase since it oxidizes cyt c at the expense of molecular oxygen. Cyt a/a3 contains copper ions as well as heme.

In facultative anaerobes there is a shorter chain and cyt c & cyt c1 are both missing. Electrons go from UQ to cyt b to cyt o or d.

In *E. coli* high aeration gives mostly the cytochrome o complex:

2 heme b (b555 & b562) and 2 copper ions.

Low aeration results in appearance of the cytochrome d complex:

2 heme b (b558 & b595) and 2 heme d, but no copper.

The Km values for O_2 are 0.2μM for cyt o and 0.02μM for cyt d which is consistent with the idea that cyt d is needed when the concentration of O_2 is low. Thes are sometimes called ubiquinol oxidases since they oxidize UQ at the expense of molecular oxygen.

There is no cytochrome c, c1 or a in the *E.coli* aerobic respiratory chain. Cyt d used to be called cyt a1 but is not really an a-type. A c-type cytochrome is found in the anaerobic nitrite reductase of *E.coli.*

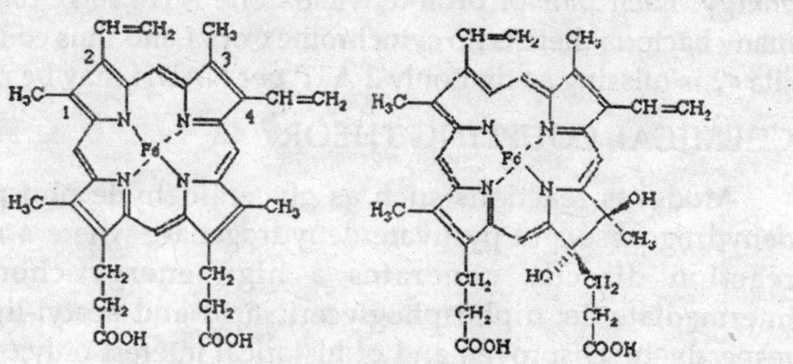

Protoheme IX = Heme b Heme d [a chlorin]

Porphyrin = Heme without the Fe atom and with 2 H atoms instead. The four 5-membered rings each containing N are pyrrole rings. Cytochromes are classified according to the substituents at positions 2 and 4 of the porphyrin/heme ring:

Cytochrome b: vinyl groups at both 2 and 4 positions. This gives protoheme IX and is also found in hemoglobin, myoglobin, catalase and peroxidase. Cytochrome o of *E. coli* is a b-type.

Cytochrome c: positions 2 and 4 both are -CH(CH_3)-S-Protein.

Cytochrome a: position 2 has a 17-carbon side chain, position 4 is vinyl and position 8 has -CHO (instead of the usual CH_3).

Chlorins are hemes in which one or more of the pyrrole rings are reduced. Heme d of *E.coli* cytochrome d (= cyt a1) and siroheme of sulfite reductase are chlorin derivatives.

THE PROTON MOTIVE FORCE

Originally, it was thought that ATP was made by direct chemical phosphorylation at 3 places along the ETC where sufficient energy was available from the drop in redox potential.

These are the three coupling sites. It is now known that 2 protons are expelled at each coupling site, so generating the Proton Motive Force (PMF).

ATP is made indirectly using the PMF as a source of energy. Each pair of protons yields one ATP. In *E. coli* and many bacteria there is no cytochrome c or c1 and thus coupling site #2 is missing so that only 2 ATP per NADH may be made.

CHEMICAL COUPLING THEORY

Model is reactions such as glyceraldehyde phosphate dehydrogenase and pyruvate dehydrogenase where a redox reaction directly generates a high energy chemical intermediate, ie: diphosphoglyceric acid and acetyl-lipoate respectively. Disproven and of historical interest only.

CHEMIOSMOTIC THEORY

ATP synthesis is coupled indirectly to electron transport via the Proton Motive Force (PMF). In the chemiosmotic model, each of the coupling sites is responsible for extruding protons, so creating a proton concentration gradient across the membrane. The ATP synthase uses this gradient to energize the synthesis of ATP.

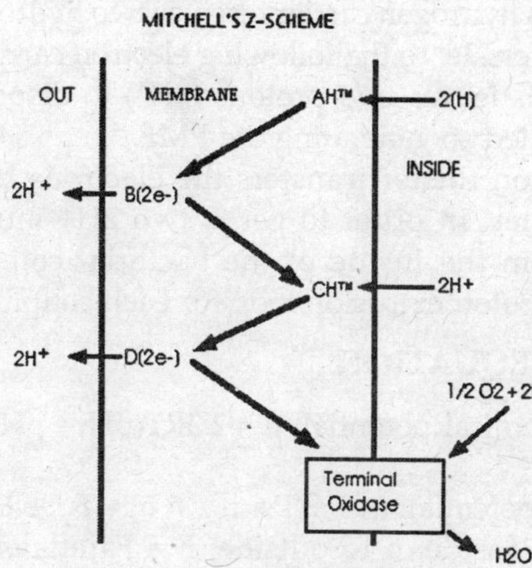

Fig. Mitchell's Z-scheme

The membrane must be impermeable to H^+ (and OH^-). Since H+ is pumped out, the exterior will become acidified and the cell interior should become alkaline relative to the outside. In practice the internal pH is held constant at 7.5 approx.

The proton gradient may be all or partially converted to a charge difference by exchanging potassium ions for protons. The PMF thus consists of two components: a pH difference (Delta pH) and a charge difference (or membrane potential, Delta Psi) across the membrane.

STOICHIOMETRY

Original theory of Mitchell has $2H^+$ extruded per site with 3 sites per mitochondrion or 2 sites per *E. coli* and hence P:O ratios of 3:1 or 2:1 assuming $2H^+$ can drive formation of one ATP. (P:O ratio = ATP per oxygen atom consumed = ATP per 2 electrons sent down ETC). Some data suggests 3 or 4 protons may be extruded at certain sites, but this only applies to mitochondria

MITCHELL'S Z-SCHEME

In Mitchell's original Z-scheme the ETC is arranged in a series of loops with hydrogen carriers facing the inside of the bacterial cell or mitochondrion and electron carriers facing outwards. Each hydrogen carrier receives two hydrogen atoms –2 and it transfers $2e^-$ to the following electron carrier. Since 2 = $2H^+$ + 2e- this leaves two protons ($2H^+$) to dispose of and these are expelled so generating the PMF.

The electron carrier transfers the electrons to the next hydrogen carrier. In order to get 2, two 2H+ must also be picked up from the inside of the bacterial cell. The ETC contains one proton extrusion loop for each coupling site.

MITCHELLS EQUATION

Proton chemical potential: μ = 2.3RT(pHi - pHo) = 2.3RT Delta pH

Electrical potential: m = F(Psi i- Psi o) = F DeltaPsi

i refers to inside, o to outside, F = Faradays constant, Psi = charge

Electrochemical potential = m(chemical) + m(electrical):

μ(H+) = F DeltaPsi - 2.3RT Delta pH.

Proton Motive Force:

PMF = μ(H+)/F = DeltaPsi - 60 Delta pH

PMF is the proton motive force in millivolts. Conversion factor is approx. 60mV/pH unit at 37°C. Note that the relative contributions of DeltaPsi and Delta pH to PMF vary with the pH of the growth medium. Note also that PMF can only be maintained as long as membrane forms a sealed vesicle.

To generate 100mV of PMF requires the extrusion of about 60,000 protons/cell. The PMF generated by *E. coli* is 150-200 mV. How the intracellular pH is maintained at pH7.5 is unknown. It is known that at an external pH of 7.5 the Delta pH is replaced by a DeltaPsi by swapping K+ for H+. In effect, *E. coli* is a tiny living battery. To convert ADP to ATP requires 7.3 kcal/mole energy input. This is equivalent to the energy stored in a pair of protons at a pH gradient of 3.5 pH units or at a charge difference of 240mV. Uncouplers are proton ionophores(= ion carriers). They carry H+ across membranes and consequently allow the PMF to collapse. Both the protonated and unprotonated forms can diffuse. Protons are picked up on the outside and released on the inside. eg: 2, 4-Dinitrophenol (2,4-DNP), carbonyl cyanide-p-trifluoromethoxyphenylhydrazone (FCCP) and the similar chloro compound CCCP.

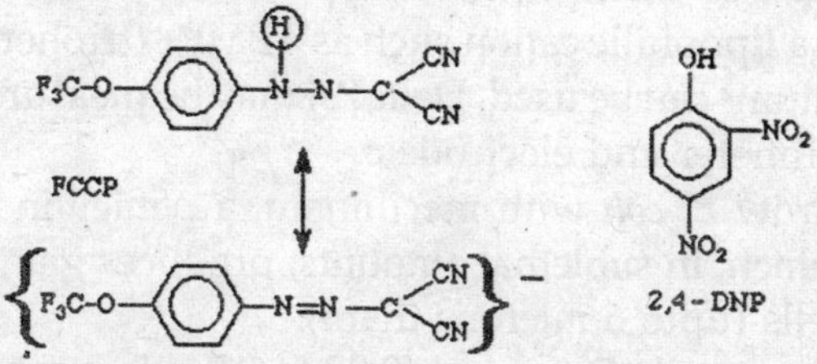

Other Ionophores

Carriers exist for ions other than the proton eg:

- Valinomycin carries potassium(K+) or rubidium(Rb+) ions and will therefore collapse the DeltaPsi but not the Delta pH
- Nigericin exchanges Na+, K+ or Rb+ for H+ so collapsing Delta pH but not DeltaPsi ie opposite effect to valinomycin
- Gramicidin is not a carrier but forms a channel in the membrane allowing through H+, Na+, K+, & Rb+ thus collapsing both Delta pH & DeltaPsi

Measurements of DeltaPsi and Delta

Delta pH may be measured by the distribution of weak acids across the membrane. This depends on the Delta pH. Acids used eg: acetic acid, salicylic acid, DMO (5,5-dimethyloxazolidine-2,4-dione):

- Membrane permeable to undissociated acid(HA) but not to H+ or A-
- On each side of membrane HA ∞ H+ + A-
- The extent of dissociation depends on pH
- Thus extent of dissociation and hence the amount of HA differs on either side of membrane and this can be measured.

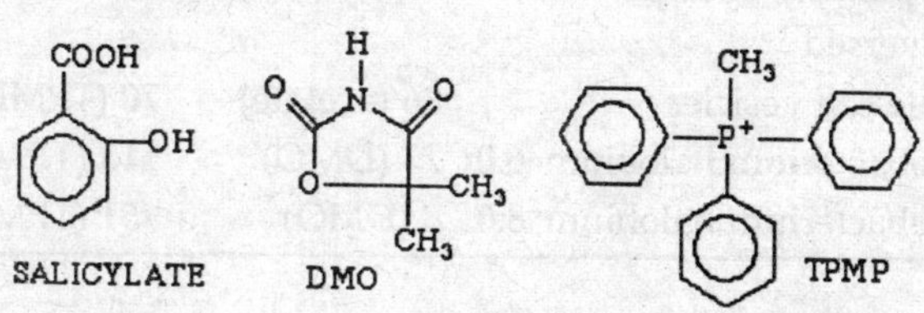

DeltaPsi may be measured indirectly by a similar method. A membrane permeable cation will be distributed according to DeltaPsi. For example, the ionophore valinomycin plus K+ or Rb+, or a lipophilic cation such as TPMP+ (triphenylmethyl-phosphonium) can be used. DeltaPsi may be measured directly with a voltmeter and electrodes:

- Grow *E. coli* with mecillinam, a penicillin derivative which, in sublethal amounts, produces giant spherical cells (upto 6 microns diam).
- Use small electrodes! (0.08 to 0.2 microns at tip).
- Coat electrodes with phospholipid to get a good seal with the cell membrane - required to avoid PMF collapsing.
- Read DeltaPsi from voltmeter.

Table. Example Proton Motive Force Values

Organism	pHo	-60DeltapH	Delta Psi	PMF
Fermentative				
Staphylococcus lactis	5.0	58 (benzoate)	95 (dye)	158
Staphylococcus faecalis	5.0	54 (DMO)	170 (DDA)	224
Fermentative/Respiratory				
Bacillus subtilis	5.0	120 (salicylate)	30 (TPMP)	150
Bacillus subtilis	7.5	0 (salicylate)	140 (TPMP)	140
Escherichia coli	5.5	72 (DMO)	57 (Rb+/Val)	129
Escherichia coli	7.0	38 (DMO)	122 (Rb+/Val)	160
Escherichia coli	5.5	120 (DMO)	100 (electrodes)	220
Escherichia coli	7.5	0 (DMO)	142 (electrodes)	142
Photosynthetic				
Rhodospirillum rubrum chromatophores* as these are inside out, polarity of PMF is reversed		72 ($CH_3NH_4^+$)	50 (SCN-)	122
whole cell vesicles		36 (acetate)	70 (TPMP)	106
Halobacterium halobium	6.0	73 (DMO)	113 (TPMP)	186
Halobacterium halobium	8.0	8 (DMO)	151 (TPMP)	159

OXIDATIVE PHOSPHORYLATION

ATP synthesis is driven by the PMF. It can be demonstrated that only the PMF and the ATP synthase enzyme complex are required in the presence of sealed membrane vesicles.

Note: Experiment is usually done "inside outΔ with the ATPase stuck to the outside.

- Make liposomes (i.e., lipid vesicles)
- Add purified ATPase and ADP and Pi
- Add excess K+ plus Valinomycin to outside
- Positive charge builds up on inside - an artificial PMF
- The artificial DeltaPsi repels protons, derived from: $H_2O = H^+ + OH^-$
- Protons exit via ATP synthase and ATP is made

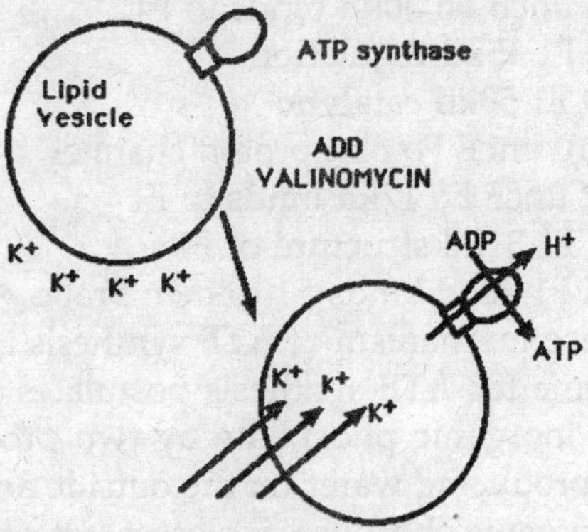

Fig. ATP synthase

ATPase and ATP synthase refer to the same enzyme - which direction the reaction goes depends on the conditions.

The ATP synthase consists of the headpiece, or F1-complex, which sticks into the interior of the cell and the Fo-complex which is membrane-bound. The catalytic a and b subunits of the F1-complex are responsible for synthesizing ATP. The Fo complex consists of 3 proteins a, b and c (sometimes referred to as c, y, w). Fo forms a proton channel through the membrane.

Protein c is very hydrophobic (a "proteolipidΔ) and

probably acts as the channel while a and b control assembly and binding to the F1-headpiece.

Fo alone makes membranes permeable to protons. Addition of F1 stops proton transport except to drive ATP synthesis. DCCD (dicyclohexyl carbodiimide) binds covalently to protein c and blocks the proton channel.

Oligomycin inhibits the ATPase reaction of the catalytic F1 complex. Oligomycin halts the whole electron transport chain & ATP synthetic pathway when added alone. However, if both oligomycin and an uncoupler are added then the ETC runs but no ATP synthesis occurs.

PROTEINS OF THE ATP SYNTHETASE

Protein Ratio Gene Subunit MW Function
α 3 uncA F1 55kd catalytic
a (or c) 1 uncB Fo 30kd binds to F1
ε 1 uncC F1 15kd regulation
β 3 uncD F1 50kd catalytic
c (or w) 10 uncE Fo 8kd proton channel
b (or y) 2 uncF Fo 17kd binds to F1
γ 1 uncG F1 31kd structure of F1
δ 1 uncH F1 20kd binds F1 to membrane

The precise mechanism of ATP synthesis is obscure. The simplest scheme for ATP synthesis postulates removal of an oxygen from inorganic phosphate by two protons from the outside thus producing water on the outside and ATP on the inside. The inorganic phosphate is closest to the proton channel with the ADP to the inside An alternative is the use of energy from the protons to alter the conformation of the catalytic a and β subunits. The high energy state then releases its energy by converting ADP to ATP.

ATP VERSUS THE FORCE

The ATPase can interconvert PMF and ATP. In the absence of a PMF the ATPase will hydrolyse ATP and generate PMF. Hence ATP and PMF are interconvertible. If an uncoupler is added to *E. coli* or a mitochondrion the PMF is dissipated. The ATPase then tries to re-create the PMF by hydrolyzing all the

ATP. Oligomycin inhibits the ATPase and will prevent hydrolysis of ATP under these conditions just as it prevents synthesis of ATP from PMF.

ATP is used for biosynthesis and also drives certain transport systems. Arsenate inhibits ATP driven transport systems, but not those energized directly by the proton motive force. In eukaryotes the PMF is generated in the mitochomdria or chloroplasts and is all converted to ATP. In bacteria the PMF is used directly to energize several membrane bound operations:-

PMF DRIVEN TRANSPORT SYSTEMS:

Symport: e.g., Lactose Substrate and H+ enter together.
Antiport: e.g., Na+ H+ taken in and Na+ excreted.
Complex: e.g., Melibiose Overall Melibiose plus H+ enter: Melibiose plus Na+ enter together ii) Na+ out/H+ in.

Energy Linked Transhydrogenase

Normally NADH is converted to ATP whereas NADPH is used as reducing power in biosynthesis. NADPH may be generated directly by certain metabolic pathways which generate NADPH instead of NADH.

A proportion of the glucose (around 20%) goes via the pentose phosphate pathway instead of the Embden-Meyerhof pathway. In many bacteria the isocitrate dehydrogenase in the Krebs cycle also makes NADPH. Another source is the NADP linked malic enzyme.

However if their is a shortage of NADPH then it is possible to convert NADH to NADPH. This requires energy in order to shift the equilibrium in favour of NADPH. Energy linked transhydrogenase is energized directly by the PMF and converts NADH + NADP to NADPH + NAD. Mechanism unknown. It is particularly important in some photosynthetic organisms which do not have any glucose to send down the pentose pathway. It is present in *E. coli* but not important.

Flagellar Motion

E. coli is powered by a proton drive - just like the ships of the galactic empire The M-ring of the basal body is located in the cytoplasmic membrane and rotates relative to the S- or stator ring of the basal body of the flagellum which is embedded in the cell wall.

Proton acceptor groups (e.g., $-NH_2$) are present on the rotor-M-ring and anionic groups (e.g., COO-) are located over the exit channels for protons.

Each time one H+ enters the input channel the ring turns a fraction of the revolution as the newly formed $-NH_3^+$ group is attracted towards the -COO- group.

ANAEROBIC RESPIRATION

Oxygen is the ultimate electron acceptor for the ETC when bacteria grow aerobically. However, bacteria can respire in the absence of oxygen if some other suitable oxidant is available to act as electron acceptor.

Use of such alternative electron acceptors is known as anaerobic respiration. It requires a modified ETC in order to transport electrons to the new acceptor. *E. coli* can use nitrate (NO_3^{-}), amine oxides (R3NO), sulfoxides (R2SO), fumarate and nitrite (NO_2).

Many sulfur compounds can be used by a variety of bacteria. "Sulfide fermentationΔ is not actually fermentation but is an example of anaerobic respiration with sulfate (SO_4^{2-}) as the electron acceptor. "Methane fermentationΔ is also anaerobic respiration, in this case with carbon dioxide (CO_2) as the electron acceptor.

Do not confuse anaerobic respiration with the oxidative metabolism of lithotrophs such as *Nitrobacter*. Oxygen has a more positive Eo than nitrite, therefore oxidation of nitrite by oxygen produces energy. Nitrate has a more positive Eo than 2 from NADH or formate.

Therefore oxidation of NADH or formate by nitrate produces energy. Lithotrophs oxidize inorganic materials using O_2. A substance with an Eo between O_2 and NADH may be used either to oxidize NADH or to reduce O_2. e.g. the NO_3^-/NO_2^- couple which has Eo of +420mV compared to +820mV for O_2/H_2O and -320mV for NAD+/NADH.

Anaerobic nitrate respiration:

$$NO_3 + 2 = NO_{2-} + H_2O$$

Nitrite oxidation by *Nitrobacter*: $NO_{2-} + 1/2O_2 = NO_{3-}$

NITRATE REDUCTASE

Nitrate reductase (NAR) of *E. coli* is a very large membrane bound protein (Total MW approx. 800,000). It is induced by nitrate and repressed by oxygen. When maximally induced it comprises upto 25% of the protein of the cytoplasmic membrane. Pyruvate formate lyase is present in anaerobic cells growing on nitrate and produces a lot of formate plus acetyl-CoA. Most of the acetyl-CoA is converted to acetate which is excreted. This is because the Krebs cycle is not complete during anaerobic growth because oxoglutarate dehydrogenase is not made. The formate is oxidized via formate dehydrogenase to give 2 plus CO_2.

The reducing equivalents travel via the cytochrome b of the FDH complex to a quinone. NADH produced in glycolysis can also be reoxidized by nitrate. The quinone used anaerobically is mostly menaquinone (Vitamin K2). However, when grown on nitrate, which is a good oxidant, a mixture of ubiquinone and menaquinone is found.

Nitrate reductase contains 4 subunits:

- NarG protein × 4 (= α) MW = 150,000, carries Mo & FeS groups
- NarH protein × 4 (= β) MW = 60,000, carries FeS groups
- NarI protein (= γ) cytochrome b, often lost on purification
- NarJ protein membrane binding subunit

The Mo is present as part of the low molecular weight (approx 1000) molybdopterin cofactor. All Mo proteins share the same cofactor, (except for nitrogenase which has a different Mo/Fe cofactor).

THE RESPIRATORY CHAIN TO NITRATE

Formate dehydrogenase is another large membrane protein. It contains molybdopterin, FeS groups and selenium. There are two types of FDH. The anaerobic respiratory FDH is found in anaerobic cells using nitrate or other alternative electron acceptors. The fermentative FDH is linked to hydrogenase and is produced anaerobically in the absence of alternative electron acceptors. The Se is present as selenocysteine ie cysteine in which sulfur has been replaced by selenium.

Most anaerobic reductases such as those for TMAO (trimethylamine oxide), DMSO (dimethyl sulfoxide) etc are molybdoproteins. Tungsten (W) is an analog of Mo. In the presence of tungstate (WO¢¤-) the processing of molybdate (MoO¢¤-) into molybdopterin is inhibited and nitrate reductase and other Mo enzymes are inactivated. Excess molybdate will remedy this i.e., tungstate is a competitive inhibitor. The two anaerobic reductases which do not contain molybdenum are nitrite reductase - a heme protein and fumarate reductase - a flavoprotein.

Fumarate Reductase (FRD) FRD is repressed by both oxygen and nitrate but induced by fumarate. FRD contains FAD and FeS groups. There are 4 subunits. A large flavoprotein, a medium sized FeS protein and two small membrane anchor proteins. FRD is connected via a cytochrome b to menaquinone. FRD cannot use ubiquinone, it has to use menaquinone. No molydenum or selenium are found in FRD. Note that fumarate is an organic molecule and the 2 go to saturate a double bond i.e., no water is formed as with nitrate, TMAO or DMSO etc.

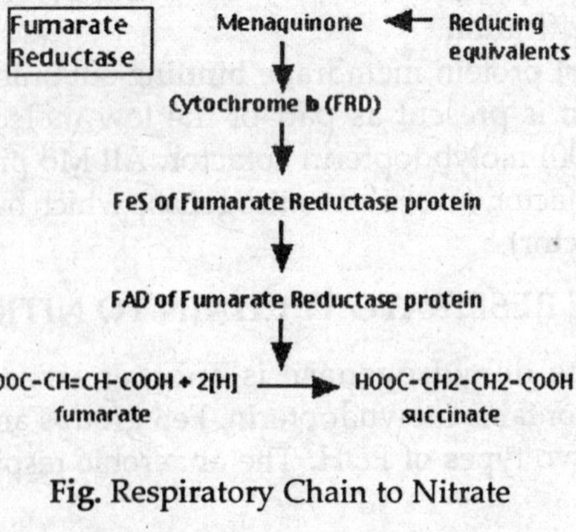

Fig. Respiratory Chain to Nitrate

Unlike O_2 and nitrate which both generate two pairs of protons per 2e- which travel down the electron transport chain and hence yield 2ATP, fumarate gives only one pair of protons and only one ATP. When a cell changes from oxygen to fumarate as electron acceptor the PMF drops from around -200mV to approx. -100mV.

OXIDATIVE PHOSPHORYLATION

The NADH and $FADH_2$ formed in glycolysis, fatty acid oxidation and the citric acid cycle are energy-rich molecules because each contains a pair of electrons having a high transfer potential. When these electrons are used to reduce molecular oxygen to water, a large amount of free energy is liberated, which can be used to generate ATP. *Oxidative phosphorylation is the process in which ATP is formed as a result of the transfer of electrons from NADH or $FADH_2$ to O_2 by a series of electron carriers.*

This process, which takes place in mitochondria, is the major source of ATP in aerobic organisms. For example, oxidative phosphorylation generates 26 of the 30 molecules of ATP that are formed when glucose is completely oxidized to CO_2 and H_2O.

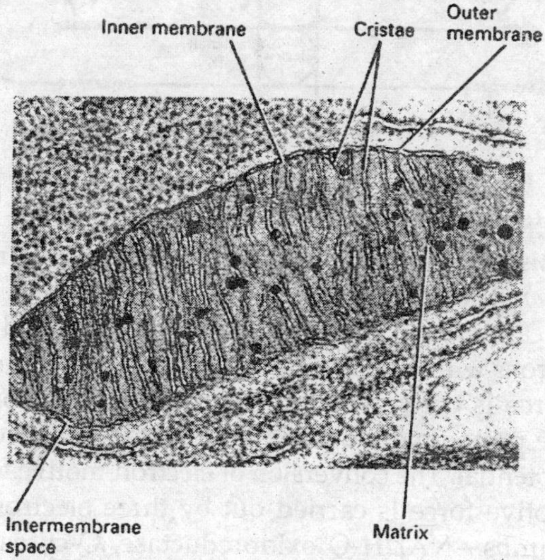

Fig. Electron Micrograph of a Mitochondrion

Oxidative phosphorylation is conceptually simple and mechanistically complex. Indeed, the unraveling of the mechanism of oxidative phosphorylation has been one of the most challenging problems of biochemistry.

The flow of electrons from NADH or $FADH_2$ to O_2 through protein complexes located in the mitochondrial inner membrane leads to the pumping of protons out of the mitochondrial matrix. The resulting uneven distribution of protons generates a pH gradient and a transmembrane electrical potential that creates a *proton-motive force.* ATP is synthesized when protons flow back to the mitochondrial matrix through an enzyme complex. Thus, *the oxidation of fuels and the phosphorylation of ADP are coupled by a proton gradient across the inner mitochondrial membrane.*

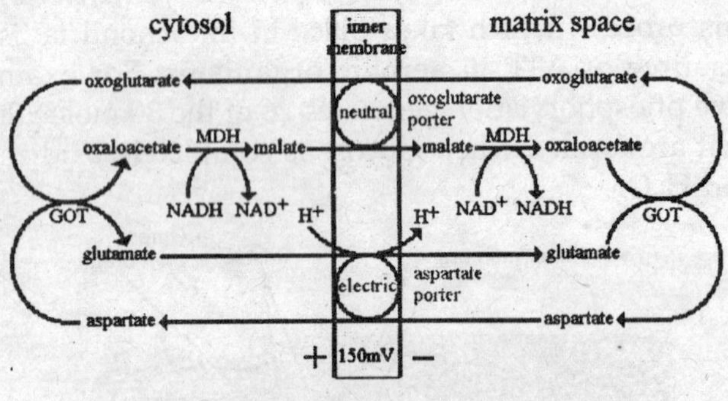

Fig. Essence of Oxidative Phosphorylation.

Oxidative phosphorylation is the culmination of a series of energy transformations that are called *cellular respiration* or simply *respiration* in their entirety.

First, carbon fuels are oxidized in the citric acid cycle to yield electrons with high transfer potential. Then, this electron-motive force is converted into a proton-motive force and, finally, the proton-motive force is converted into phosphoryl transfer potential. The conversion of electron-motive force into proton-motive force is carried out by three electron-driven proton pumps—NADH-Q oxidoreductase, Q-cytochrome *c* oxidore-ductase and cytochrome *c* oxidase.

These large transmembrane complexes contain multiple

oxidation-reduction centers, including quinones, flavins, iron-sulfur clusters, hemes and copper ions. The final phase of oxidative phosphorylation is carried out by *ATP synthase,* an ATP-synthesizing assembly that is driven by the flow of protons back into the mitochondrial matrix. Components of this remarkable enzyme rotate as part of its catalytic mechanism. Oxidative phosphorylation vividly shows that *proton gradients are an interconvertible currency of free energy in biological systems.*

THE GLYOXYLATE CYCLE

It was stated above that fatty acids cannot be utilized for gluconeogenesis. Also, it was mentioned earlier that fatty acids are degraded via acetyl-CoA. Since acetyl-CoA is degraded via the TCA and TCA intermediates can be utilized for gluconeogenesis, why is it not possible to turn fatty acids into glucose? This is because of the different roles of the TCA intermediates in the two processes:

- In the degradation of acetyl-CoA, they have a catalytic role – they are required for the cycle to run, but there is no net gain or loss of intermediates. Acetyl-CoA fed into the TCA is consumed without increasing the pool of TCA intermediates.
- In gluconeogenesis, oxaloacetate is drained from the pool of TCA intermediates. Thus, the TCA can only sustain gluconeogenesis to the extent it is supplied with new intermediates from other sources.

To turn fatty acids into glucose, we would therefore need a means to use acetyl-CoA for increasing the pool of intermediates. Such a pathway does not exist in mammals, but it does exist in plants; it is known as the glyoxylate cycle. It is a 'side-road' to the TCA that allows them to use two molecules of acetyl-CoA per cycle to bring about the net synthesis of one C_4-intermediate.

Two reactions are required for this cycle:

- Isocitrate is split into succinate and glyoxylate by isocitrate lyase. Since the isocitrate dehydrogenase and the á-ketoglutarate dehydrogenase reactions are bypassed, the loss of two carbons as CO_2 is avoided; these carbons are retained in the form of glyoxylate.

- Glyoxylate combines with an 'extra' acetyl-CoA to form one molecule of malate. This reaction is catalysed by malate synthase and like the citrate synthase reaction it is pushed forward by the concomitant hydrolysis of coenzyme A.

We know that many plant seeds are very rich in oil (=fat). The glyoxylate cycle enables plant seeds to store metabolic energy and carbon as fat and to use it for the synthesis of glucose and other carbohydrates during germination.

Fat is water-free and has approximately twice the content of metabolic energy per gram of carbohydrates; it is therefore both more compact and more resistant to microbial degradation than starch.

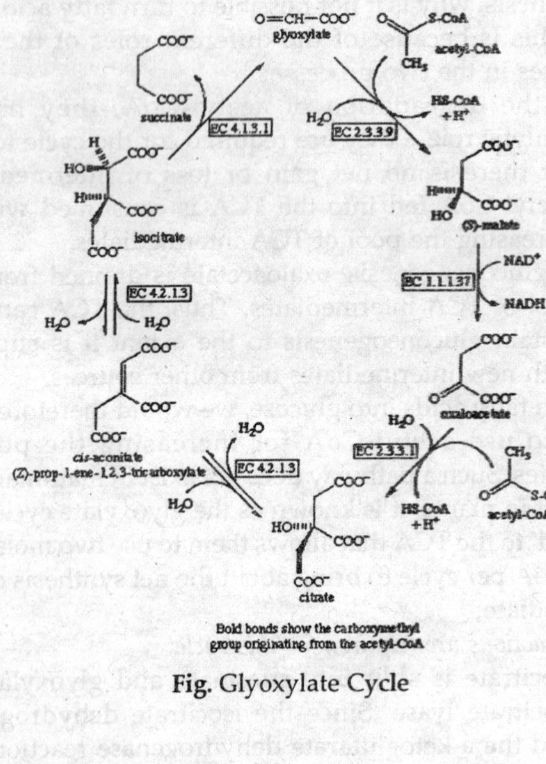

Fig. Glyoxylate Cycle

Chapter 4

Vitamins Biosynthesis

Vitamin, any of the organic (carbon-containing) compounds that the body requires in small amounts to maintain health and function proper'y. Children additionally need vitamins to grow. The body gets most of its vitamins from the foods we eat.

A healthy diet with plenty of fruits and vegetables should provide nearly all of the vitamins a person needs. Scientists have classified 13 compounds as vitamins. They have given most of these vitamins letter or letter plus number names, such as A, B12, and D. Most vitamins are produced by plants. Some, such as vitamin D, are produced only by animals. A few vitamins are made by the body itself. For example, bacteria in the digestive tract help produce vitamin K, and the skin uses sunlight to produce vitamin D.

Vitamins are also manufactured for sale as supplements for people who need additional vitamins to meet their body's requirements. For example, doctors often prescribe vitamin supplements for pregnant and nursing women to provide the additional nutrients needed by a rapidly growing fetus or infant. Older adults may not meet their vitamin requirements through food because the body's ability to absorb vitamins is impaired with age.

Expressed in milligrams (mg) or international units (IU) for adults and children of normal health, these recommendations provide useful guidelines for daily intake. Such guidelines are useful not only for professionals in nutrition but also for the growing number of families and individuals who eat irregular meals and rely on prepared foods, most of which are now required to carry nutritional labeling.

VITAMINS ARE VITAL

The word vitamine, later shortened to vitamin, was coined by Polish-American chemist Casimir Funk in the early 20th century. Funk was searching for the then-unknown substance in foods that prevents such diseases as beriberi, rickets, and scurvy. After experiments on pigeons, Funk guessed that an amine-a compound containing nitrogen and hydrogen-in foods was responsible. He called this substance a vital (necessary for life) amine, or vitamine. Scientists later discovered that not all vitamins were amines, however.

Vitamins help the body carry out essential biochemical processes. Vitamins generally combine with proteins to create enzymes that promote chemical reactions. These enzymes play an important role in metabolism-the reactions that break down the fats, carbohydrates, and proteins in food so that the body can use them for energy and cell repair.

The enzymes also promote reactions involved in the formation of bone, hormones, blood cells, nervous-system chemicals, and genetic material. Without vitamins, many of these reactions would slow down or cease. The different vitamins are not chemically related, and most differ in their actions in the body.

Deficiency of particular vitamins can lead to various diseases. Too little vitamin C, for example, can cause rickets, a disease in which the bones fail to develop properly. Bowlegs or"knock knees" are signs of rickets. Rickets and the other vitamin-deficiency diseases that interested Casimir Funk have largely disappeared in the developed world as the result of fortified foods and improved nutrition. But these diseases still occur in the developing world, especially where malnutrition is common.

Interest in vitamins grew during the 1990s, especially in those vitamins that act as antioxidants-vitamins A, C, and E. Antioxidants neutralize molecules known free radicals that cause cell damage. Free radicals have been linked to a number of disorders, including Alzheimer's disease, arteriosclerosis, cancer, diabetes, and Parkinson disease. Researchers theorized that increased doses of antioxidants might also prevent and even cure such diseases.

These claims have so far not panned out, however, as scientific studies have failed to demonstrate the preventive abilities of antioxidants. Research on antioxidants and other vitamins continues.

WATER-SOLUBLE VITAMINS

Water-soluble vitamins include eight well-recognized members of the vitamin B complex: thiamine, riboflavin, pyridoxine, pantothenic acid, niacin, biotin, folic acid and vitamin B12; the water-soluble essential nutritional factors: choline, inositol, ascorbic acid; and vitamins with less-defined activity for fish: p-aminobenzoic acid, lipoic acid and citrin.

Table. Water-Soluble Vitamin Test Diet H-440

Complete test diet (g)		Vitamin mix (mg)		Mineral mix (mg)	
Vitamin-free casein	38	Thiamine HCl	5	USP XII No. 2	plus
Gelatin	12	Riboflavin	20	$AlCl_3$	15
Corn oil	6	Pyridoxine HCl	5	$ZnSO_4$	300
Cod liver oil	3	Choline chloride	500	CuCl	10
White dextrin	28	Nicotinic acid	75	$MnSO_4$	80
α-Cellulose mixture	9	Calcium pantothenate	50	KI	15
α-Cellulose	8	Inositol	200	$CaCl_2$	100
Vitamins		Biotin	0.5	per 100 g of salt mixture	
		Folic acid	1.5		
Mineral mix	4	L-Ascorbic acid	100		
Water	200	Vitamin B12	0.01		
Total diet as fed	300	Menadione (K)	4		
		-Tocopherol acetate (E)	40		

The first eight, though required only in small amounts in the diet, play major roles in growth, physiology and metabolism. Choline, inositol and ascorbic acid are required in appreciable quantities in the diet and sometimes are not referred to as vitamins but as major dietary nutrients. A typical vitamin test diet for fish control over the water-soluble vitamins

THIAMINE

Experimental avian polyneuritis, a condition resembling beriberi, was produced by Eijkman in Java in 1886 and the anti-beriberi factor was crystallized and named"vitamine". The term"thiamine" was introduced when the chemical nature of the factor was established. Thiamine was isolated from rice polishings in 1926 and synthesized in 1936.

Thiamine hydrochloride is a water-soluble, colourless, monoclinic, crystalline compound. It is comparatively stable to dry heat but is rapidly broken down in neutral or alkaline solutions and is split by sulphites into constituent pyrimidine and thiazole moieties.

It has a characteristic yeast-like odour. The pyrimide ring is relatively stable, but the thiazole ring is easily opened by hydrolysis.

Several derivatives are stable to heat and appear to be more completely soluble in weak alkaline solutions than thiamine itself and still show biological activity in animals.

These derivatives include thiamine propyl disulphide, benzoylthiamine disulphide, dibenzoylthiamine, and benzoylthiamine monophosphate. Both thiamine hydrochloride and thiamine mononitrate have been successfully used as the active vitamin in test diets for fish nutrition studies.

POSITIVE FUNCTIONS

Thiamine functions in all cells as the coenzyme cocarboxylase, thiamine pyrophosphate, which participates in the oxidative decarboxylation of pyruvic acid to acetate for entry into the tricarboxylic acid (TCA) cycle.

THIAMINE PYROPHOSPHATE

Thiamine pyrophosphate is also a coenzyme of the

transketolase system by which direct oxidation of glucose occurs in the cytoplasm of cells via the pentose phosphate pathway. Erythrocyte (RBC) levels of metabolites of this system have been used to indicate thiamine status in experimental animals, including salmon and trout.

Thiamine is essential for good appetite, normal digestion, growth and fertility. It is needed for normal functioning of the nervous tissue and the requirement is determined by the caloric density of the diet.

Deficiency Syndrome

Deficiency signs in salmonids include impaired carbohydrate metabolism, nervous disorders, poor appetite, poor growth, and increased sensitivity to shock. A trunk-winding symptom in eels has been reported, together with haemorrhage at the base of the fins.

Skin congestion and subcutaneous haemorrhage occurs in carp fed thiamine-deficient diets. Typical symptoms observed in salmonids, carp and catfish. Thiamine deficiency has also been reported in marine flatfish started on clam neck diets stored long enough for thiaminase present to hydrolyze the thiamine in the ration. Typical nervous paralysis occurred with rapid mortality from physical shock.

Table. Vitamin Deficiency Syndromes

Vitamin	Symptoms in salmon, trout, carp, catfish
Thiamine	Poor appetite, muscle atrophy, convulsions, instability and loss of equilibrium, oedema, poor growth
Riboflavin	Corneal vascularization, cloudy lens, haemorrhagic eyes - photophobia, dim vision, incoordination, abnormal pigmentation of iris, striated constrictions of abdominal wall, dark colouration, poor appetite, anaemia, poor growth
Pyridoxine	Nervous disorders, epileptiform fits, hyperirritability, ataxia, anaemia, loss of

	appetite, oedema of peritoneal cavity, colourless serous fluid, rapid postmortem rigor mortis, rapid and gasping breathing, flexing of opercles
Pantothenic acid	Clubbed gills, prostration, loss of appetite, necrosis and scarring, cellular atrophy, gill exudate, sluggishness, poor growth
Inositol	Poor growth, distended stomach, increased gastric emptying time, skin lesions
Biotin	Loss of appetite, lesions in colon, colouration, muscle atrophy, spastic convulsions, fragmentation of erythrocytes, skin lesions, poor growth
Folic acid	Poor growth, lethargy, fragility of caudal fin, dark colouration, macrocytic anaemia
Choline	Poor growth, poor food conversion, haemorrhagic kidney and intestine
Nicotinic acid	Loss of appetite, lesions in colon, jerky or difficult motion, weakness, oedema of stomach and colon, muscle spasms while resting, poor growth*Vitamin Symptoms in salmon, trout, carp, catfish*
Vitamin B12	Poor appetite, low haemoglobin, fragmention of erythrocytes, macrocytic anaemia
Ascorbic acid	Scoliosis, lordosis, impaired collagen formation, altered cartilage, eye lesions, haemorrhagic skin, liver, kidney, intestine, and muscle
p-Aminobenzoic acid	No abnormal indication in growthap-petite, mortality

Requirements

In determining dietary thiamine requirements of fish, some considerations must be placed upon the composition of dietary ingredients in the ration.

Fat content of the diet may affect not only caloric intake but also the thiamine requirement because cocarboxylase participates in the oxidation of that through a-ketoglutarate. Therefore, fish on a high fat diet and low thiamine intake might take longer to develop deficiencies and will give an erroneous requirement.

Sources and Protection

Common sources for thiamine are dried peas, beans, cereal bran, and dried yeast. Fresh glandular tissue is also a good source for thiamine and other members of the vitamin B water-soluble complex, but is seldom used in modern commercial fish diets.

Thiamine can be easily lost by holding diet ingredients too long in storage or by preparing the diet under slightly alkaline conditions or in the presence of sulphide. Wet or frozen diets pose a problem because moisture content increases the chance of enzymatic (thiaminase) hydrolysis and subsequent destruction of thiamine. Obviously, wet or moist diet, preparations containing any fresh fish or shellfish tissue must be used immediately.

Antimetabolites and Inactivation

Acetylcholine is an antagonist to thiamine and pyrithiamine. Oxythiamine and normal butylthiamine are specific antimetabolites. Several thiaminases occur which destroy thiamine.

These rupture the thiazole ring at the sulphur bond making the residue inactive. Freshwater fish tissues have high thiaminase activity as do tissues from clams, shrimp and mussels. Thiaminases have also been found in beans and mustard seed and in several micro-organisms.

Thiaminase activity is low in most saltwater fish tissues, however, and the enzyme is inactivated by heating or prolonged pasteurization. Thiamine present in fresh Torula yeast is relatively unavailable to fish, but the yeast becomes an excellent thiamine source after rupturing the cells by steam treatment or by dehydration.

Clinical Assessment

Clinical assessment of thiamine status have been made by measuring erythrocyte transketolase activity. Typical saturated levels for thiamine activity in sea salmon range from 15 to 20 g of thiamine/g of wet liver tissue. Fingerling chinook or coho salmon reared in 10 or 15 C water on test diets containing the listed amount or more of thiamine hydrochloride: assayed at 8-10 g of thiamine/g of wet liver tissue.

These liver storage levels and normal erythrocyte transketolase activity in the absence of any deficiency sign will indicate an adequate thiamine intake for that fish population.

RIBOFLAVIN

Following the recognition that the original water-soluble B consisted of more than one factor, a new growth-promoting factor consisting of yellow-green pigments was isolated in 1879. This new factor was designated G by some and 83 by others.

The second water-soluble vitamin discovered was given its specific chemical name riboflavin. Lactoflavin, hepatoflavin, and ovoflavin were also shown to be identical with the pure riboflavin. Riboflavin occurs in the free form only in the eye, whey and urine.

Table. Vitamin Requirements for Growth

Vitamin (mg/kg dry diet)	*Rainbow trout*	*Brook trout*	*Brown trout*	*Atlantic salmon*	*Chinook salmon*	*Coho salmon*
Thiamine	10-12	10-12	10-12	10-15	10-15	10-15
Riboflavin	20-30	20-30	20-30	5-10	20-25	20-25
Pyridoxine	10-15	10-15	10-15	10-15	15-20	15-20
Pantothenate	40-50	40-50	40-50	*	40-50	40-50
Niacin	120-150	120-150	120-150	*	150-200	150-200
Folacin	6-10	6-10	6-10	5-10	6-10	6-10
Cyanocobalamin	*	*	*	*	0.015-0.02	0.015-0.02
myo-Inositol	200-300	*	*	*	300-400	300-400
Choline	*	*	*	*	600-800	600-800

Biotin	1-1.5	1-1.5	1.5-2	-	1-1.5	1-1.5
Ascorbate	100-150	*	*	*	100-150	50-80
Vitamin A	2 000-2 500 I.U.	*	*	-	*	*
Vitamin E	*	*	*	-	40-50	*
Vitamin K	*	*	*	-	*	*

*Denotes a requirement, the level of which has not been established

Table. Vitamin Requirements for Growth

Vitamin (mg/kg dry diet)	Carp	Channel catfish	Eel	Sea bream	Turbot	Yellowtail
Thiamine	2-3	1-3	2-5	*	2-4	*
Riboflavin	7-10	*	*	*	*	-
Pyridoxine	5-10	*	*	2-5	*	*
Pantothenate	30-40	25-30	*	*	*	*
Niacin	30-50	*	-	*	-	-
Folacin	-	*	*	-	*	-
Cyanocobalamin	-	*	-	*	-	-
myo-Inositol	200-300	*	-	300-500	-	-
Choline	500-600	*	-	*	*	-
Biotin	1-15	*	*		*	*
Ascorbate	30-50	30-50	-	*	-	*
Vitamin A	1 000-2 000 I.U.	*	-	-	-	*
Vitamin E	80-100	*	-	-	-	*
Vitamin K	*	*	-	-	-	-

*Denotes a requirement, the level of which has not been established

RIBOFLAVIN

Riboflavin is a yellow-brown crystalline pigment. The vitamin is very slightly soluble in water but soluble in alkali. It is insoluble in most organic solvents. Riboflavin is stable to oxidizing agents in strong mineral acids and in neutral aqueous solution. It is also stable to dry heat but is irreversibly decomposed on irradiation with ultraviolet rays or visible light, breaking down to lumiflavin. Riboflavin phosphate is the chemically active group of Warburg's'yellow' enzyme.

POSITIVE FUNCTIONS

Riboflavin functions in the tissues in the form of flavin adenine dinucleotide (FAD) or as flavin mononucleotide (FMN). The flavo-proteins function as enzymes of tissue respiration and are involved in hydrogen transport to catalyze the oxidation of reduced pyridine nucleo-tides (NADH and NADPH).

Thus, they function as coenzymes for many oxidases and redunctases such as cytochrome e reductase, D- and L-amino acid oxidases, xanthine and aldehyde oxidase, succinic dehydrogenase, glucose oxidase and fumaric dehydrogenase. Riboflavin is also involved with pyridoxine in the conversion of tryptophan to nicotinic acid and is most important in the respiration of poorly vascularized tissues such as the cornea of the eye.

Riboflavin is involved in the retinal pigment during light adaptation and lack of it causes impaired vision and photophobia in experimental animals, including fish.

DEFICIENCY SYNDROME

Tissue riboflavin is exhausted in young salmonids after 10 to 12 weeks on riboflavin-deficient diets in 10-15 °C water systems. Poor appetite and poor diet efficiency are the first signs, followed by photophobia, mono or bilateral cataract's, corneal vascularization, eye haemorrhage, incoordination, and general anaemia.

Dark pigmentation, coupled with striated constrictions of abdominal wall in salmon have been noted. Skin atrophy has been reported for some fish species and abnormal pigmentation of both skin and iris has been noted. Replacement of riboflavin in the diet reduced the symptoms except when cataracts have developed.

This irreversible condition will continue in monolateral cataracts throughout the life of the fish, whereas bilateral cataracts largely result in eventual starvation and death of the afflicted animal. The first specific signs have consistently appeared in and about the eyes of salmonids, carps and catfish.

REQUIREMENTS

The requirements of fish for riboflavin under experimental conditions in 10-15°C water supplies Values for the trout are slightly lower than those reported for salmon. The requirements may vary depending upon the balance of other dietary ingredients, caloric density and environmental conditions under which the fish is raised.

The values established under these standard test conditions should serve as tentative-levels which will satisfy biological demands for normal growth, health, and physiological function. Most of these studies have been made on very young fish, often initially feeding fry, with the logical assumption that their vitamin needs would be more than that of larger fish having advanced metabolic enzyme systems capable of synthesizing at least some of these vitamins to partially meet the requirements.

SOURCES AND PROTECTION

Riboflavin is widely distributed in plants and in animal glandular tissues. Milk, liver, kidney, heart, yeast, germinated grains, peanuts, soybeans and eggs are rich sources. Keeping feeds from sunlight or intense artificial light is necessary to minimize loss of the vitamin by conversion to lumiflavin.

ANTIMTABOLITES AND INACTIVATION

When the ribose group in the molecule is replaced by other groups, analogues are formed which either have reduced activity or become antimetabolites. Galactoflavin is an antagonist to riboflavin and inhibits growth of rats when the diet contains this compound. Flavin monosulphate inhibits D-amino acid oxidase and appears to act as a competitor and inhibitor for growth of Lactobacillus casei. The hydroxyethyl analogue is an antagonist for riboflavin function in both rats and bacteria and also shows antifungal activity.

CLINICAL ASSESSMENT

Liver tissue of actively feeding sea salmon contains 6-8 g of riboflavin/g of wet tissue. In a freshwater environment, young fish fed test diets in 15 ºC water systems showed liver storage of 3.5-4.0 ωg/g. Riboflavin content of blood plasma does not change significantly in riboflavin deficiency in experimental animals. However, the erythrocyte riboflavin content has been reported at around 10 g/100 ml blood for man on low riboflavin intake and approximately twice that level on high intake.

Urinary excretion of riboflavin has been used clinically to indicate dietary status, with 200ωg/24 it suggesting adequate intake and less than 100ωg indicating low intake. Excretion of 50ωg or less daily is strong indication of extended dietary deficiency. Several studies show that excretion of less than 200ωg of riboflavin/g of creatinine would be indicative of deficiency.

PYRIDOXINE

A new factor which could cure dermatitis in rats was reported by Györgi in 1935 and was named vitamin B6. The name pyridoxine is also attributed to Györgi. The active material was isolated in 1939 and pyridoxine was subsequently synthesized. First quantitative requirements for fish were described in 1944. Pyridoxine deficiency has since been reported in trout and in salmon.

Pyrodoxine Hydrochloride

Compounds which have vitamin B6 activity include: pyridoxine, pyridoxal and pyridoxamine. Pyridoxine hydrochloride is readily soluble in water and is, heat stable in either acid or alkaline solution. It acts as a coenzyme in a number of enzyme systems. Pyridoxine is sensitive to ultraviolet light in neutral or alkaline solutions.

Pyridoxamine and pyridoxal in dilute solutions are labile compounds which are rapidly destroyed on exposure to air, heat, or light. Therefore, most vitamin supplementation is in the form of pyridoxine hydrochloride and analysis for pyridoxine activity by microbiological assay of diet ingredients probably measures pyridoxal phosphate and other intermediates as well.

Positive Functions

Pyridoxal phosphate is the coenzyme, codecarboxylase, involved in the decarboxylation of amino acids. It is also the co-factor of the 22 different transaminases present in animal tissues. As codecarboxylase, pyridoxal phosphate has been shown to participate in the decarboxylation of 5-hydroxytryptophan to produce 5-hydroxytryptamine or serotonin. As a co-factor of the enzyme desulphydrase converts cysteine to pyruvic acid.

Pyridoxal phosphate is the co-factor for the synthesis of -amino-levulinate, the latter a precursor of haeme. Many neuro hormones require pyridoxal phosphate as a coenzyme in their synthesis. Pyridoxine is also involved in fat metabolism, especially of the essential fatty acids. It is involved in the synthesis of messenger RNA which determines amino acids sequence in polypeptide synthesis. Pyridoxine plays a most important role in protein metabolism and, as a result, carnivorous fish have stringent requirements for the vitamin in the diet and stores are rapidly exhausted.

Deficiency Syndrome

Signs of pyridoxine deficiency in fish Since salmonids, ictalurids, and very young cyprinids are carnivorous, with protein requirements for young animals between 40-50 per cent of the ration, pyridoxine stores are rapidly exhausted when fish are held on pyridoxine-deficient rations. Acute deficiency signs occur in salmon after 14-21 days on a high protein diet of pyridoxine and the entire population dies within 28 days in 12-15 C water.

Deficiency is followed by general nervous disorders, and alteration in control of melanophore contraction. Postmortem rigor mortis occurs very rapidly. Rapid and gasping breathing with flexing of the opercules is a common observation and oedema in the peritoneal cavity with colourless serous fluid occurs in fish on some experimental treatments. Salmon, trout, carp and yellowtail exhibit premortem rigor a few hours before death. Recovery at late stage deficiency is very unlikely unless fish are injected with pyridoxal phosphate.

Handling the animals generally induces more damage than vitamin administration will correct. Recovery among fish still feeding is equally rapid and dramatic upon administration of pyridoxine hydrochloride in the diet. Deficiency signs disappear within a day or two. Erythrocyte and plasma transaminase activities reflect the deficiency state.

Requirements

Pyridoxine requirements for trout, salmon, carp, and yellowtail Supplementing commercial fish diets with pyridoxine is desirable because of the important role the vitamin plays in protein metabolism.

Sources and Protection

Good sources of pyridoxine are yeast, whole cereals, egg yolk, liver and glandular tissues. Pyridoxine compounds in phosphorylated form present in agricultural products are fairly stable but are sensitive to ultraviolet radiation. Some pyridoxal phosphate will be lost on exposure to air. Free forms or pyridoxal and pyridoxamine are rapidly destroyed by air, light and heat when moist.

Antimetabolites and Inactivation

Antagonists may compete for reaction sites of the apoenzyme or may react with pyridoxal phosphate to form inactive compounds. Deoxypyridoxine is a potent pyridoxine antagonist because of competition for apoenzyme sites but is a useful agent to accelerate pyridoxine deficiency in experimental animals. This same compound inhibits tyrosine decarboxylase. Methoxypyridoxine is another antagonist. Oxopyrimidine (2-methyl-4-amino-5-hydroxymethyl-pyrimidine) produces liver damage and inhibits glutamate decarboxylase.

Clinical Assessment

Plasma and erythrocyte transaminase activity in most instances reflects pyridoxine status of fish although high tryptophan load in the diet tends to increase the pyridoxine requirement. Liver storage measured by microbiological assay showed 5-6 V-g of pyridoxine activity/g in fresh sea salmon liver; whereas, fingerling salmon fed a 50 per cent protein diet in freshwater had 2-3 mg/g of wet tissue.

Assay of diet for pyridoxine by microbiological methods gives a truer representation of total pyridoxine activity including intermediates.

Pantothenic Acid

A chick dermatitis was cured by Elvehjem and Köehn in 1935 using a factor containing -alanine. The term pantothenic acid was given earlier by Williams to a yeast growth factor which was later recognized as the anti-dermatitis factor. Pantothenic acid was synthesized by Stiller in 1940. Phillips observed clubbed gills in trout fed pantothenic acid-deficient diets in 1945. Rucker detected the same condition in salmon fed low pantothenic acid diets. Pantothenic acid may be considered as dihydroxydimethylbutyric acid bonded to -alanine.

The free acid is a yellow, viscous oil and therefore the compound generally used in fish diet preparation is the calcium salt. This salt is a white crystalline powder readily soluble in water, mild acid, and is almost insoluble in organic solvents. It is stable to oxidizing and reducing agents and to autoclaving, but is labile to dry heat, hot alkali, or hot acid. Pantothenol has almost as much activity as pantothenic acid for growth of chicks.

Pantothenic acid acetate, benzoate, and diphosphate esters are biologically active for animals but not for lactic acid bacteria. The optical isomer L-pantothenic acid, appears physiologically inert. Some organisms may utilize a portion of the molecule. Bacteria appeared to require only the dihydroxydimethyl-butyric acid, and some yeasts utilize only g-alanine. Animals, however, need the entire pantothenic acid molecule or its reduced alcohol form.

Positive Functions

Pantothenic acid is part of acetyl coenzyme A which

occurs in many enzymatic processes involving 2-carbon compounds. It has been shown to be required by all animal species studied and by many micro-organisms.

The acetyl coenzyme A system is involved in the acetylation of aromatic amines and choline; condensation reactions for synthesis of acetate, fatty acids, and citrate; the oxidation of pyruvate and acetaldehyde; and is essential for the development of the central nervous system.

The 2-carbon fragment called'active acetate', or acetyl coenzyme A, is an essential intermediate in metabolism. It is involved in acylation of acetate, succinate, benzoate, propionate and butyrate. Pantothenic acid is involved in adrenal function and for the production of cholesterol. Coenzyme A is also involved in many other steps of intermediate metabolism of carbohydrates, fats and proteins.

Deficiency Syndromes

Deficiency signs for pantothenic acid Under standard test conditions with deficient diet fed in 10-15°C water systems, salmon and trout exhaust pantothenic acid stores within 8-12 weeks.

Fish stop feeding and close examination of gill filament show proliferation of epithelial surface plus swelling and clubbing together of the filaments and lamellae. The opercules become distended and the surface of the gills is often covered with an exudate. Fish become prostrate or sluggish. Necrosis, scarring and cellular atrophy of the tender gill elements occur and anaemia develops after prolonged deficiency.

The same type of symptom has been observed in salmon, trout, eel, carp and catfish. After replacement of pantothenic acid in the diet, recovery is rapid for those fish still feeding and gross deficiency symptoms disappear after about four weeks on the recovery diet although evidence of necrosis and scarring remains.

Sources and Protection

Good sources for pantothenic acid are cereal bran, yeast, liver, kidney, heart, spleen and lung. Fish flesh is a relatively rich source, although the content is only about 20 per cent of pantothenic acid found in animal glandular tissue. Royal jelly probably contains the greatest amount as it contains over 500ωg of pantothenol/g dry weight.

The calcium or sodium salt of pantothenic acid is relatively stable and can be incorporated into either moist or dry fish diets. Some loss is incurred during autoclaving and excessive heat should therefore be minimized during diet preparations. Since the free acid is labile to heat and also to acid and alkali, some loss can be expected during moist diet preparation or during storage.

Certain cereal brans may have pantothenic acid bound in a form unavailable to fish because of the low digestibility coefficient and should not be relied upon as the sole pantothenic acid source in the diet.

Antimetabolites and Inactivation

Since pantothenic acid affects the respiration of many types of cells, compounds like 6-mercaptopurine, 2,6-diaminopurine, and 8-azaguanine, which inhibit growth of tumors are antagonistic toward pantothenic acid.

Pantoyltaurine is an antimetabolite of pantothenic acid and has been used to accelerate deficiency syndromes in experimental animals. Methyl- -pantothenic acid was reported to interfere with the formation of acetyl coenzyme A and accelerates deficiency syndromes in animals.

This compound inhibits sulfanilamide acetylation in pigeon liver homogenates but did not prevent citric acid formation. Pantothenic acid can be used to overcome these inhibitory effects, including the reversal of mitosis blockage by 7-mecaptopurine in animals. High levels of calcium pantothenate in the diet also has a transient effect on the migratory urge of salmon.

Clinical Assessment

Assay of pantothenic acid content in the diet may be misleading unless care is exercised in proper hydrolysis of the raw materials being assayed since pantothenic acid is only slowly liberated by normal hydrolytic procedures. Complete hydrolysis with enzyme preparations will liberate all of pantothenic acid from biologically active material of glandular tissues, fish flesh, yeast, and bran.

Salmon feeding actively in the oceans showed liver pantothenic acid content of 18-20/g of fresh tissue. Young chinook and coho salmon fingerlings reared in fresh water at 12-15°C showed maximum liver storage at about 14-16ωg of pantothenic acid/g of fresh liver tissue. Load tests and acetylation reactions of sulfanilamide have been used in man and other experimental animals but have not yet been extended for assessment of pantothenic acid status in fish.

NIACIN

Nicotinic acid was synthesized in 1873 but was left on the shelf as an organic compound unrelated to the severe pellagra afflictions occurring throughout the world at that time. Sixty years later the compound was shown to be present in coenzymes I and II and two years thereafter Elvehjem cured'black tongue' in dogs with the vitamin.

Nicotinic Acid

Niacin is the preferred nomenclature. Nicotinic acid amide or niacinamide is the common form in which the vitamin is physiologically active. Niacin is a white, crystalline solid, soluble in water and alcohol.

It is stable in the dry state and may be autoclaved for short periods without destruction. It is also stable to heat in mineral acids and alkali. Niacin is both a carboxylic acid and an amine and forms quaternary ammonium compounds because of its basic nature. Acidic characteristics include salt formation with alkali. Niacin can be esterified easily, then converted to amides.

Niacinamide is a crystalline powder soluble in water and ethanol and the dry material is stable up to about 60°C. In aqueous solutions it is stable for a short period when autoclaved. It is the form in which the vitamin is normally found in niacinamide adenine dinucleotide (NAD) and in niacinamide adenine dinucleotide phosphate (NADP).

POSITIVE FUNCTIONS

The major function of niacin in NAD and NADP is hydrogen transport in intermediary metabolism. Most of these enzyme systems function by alternating between the oxidized and reduced state of the coenzymes NAD-NADH and NADP-NADPH.

Oxidation-reduction reactions may be anaerobic as when pyruvate acts as the hydrogen acceptor and lactate is formed, or the reactions may be coupled to electron transport systems with oxygen as the ultimate hydrogen acceptor; such aerobic reactions occur in respiration. Both NAD and NADP are involved in the synthesis of high energy phosphate bonds which furnish energy for certain steps in glycolysis, in pyruvate metabolism, amino acid and protein metabolism, and in photosynthesis.

An interrelationship between thiamine and niacin exists since both vitamins are coenzymes in intermediary metabolism whereby food is oxidized to furnish energy for physiological functions, to maintain homeostasis, and for body temperature maintenance in homeotherms.

DEFICIENCY SYNDROME

Stores of niacin are more slowly exhausted under experimental conditions than are some of the other vitamins resulting in less defined and more slowly developing symptoms. Niacin deficiencies in fish were experimentally induced in the late forties and early fifties by using basal diets which had a low niacin content. Loss of appetite and poor food conversion were the first signs noted.

Then the fish turned dark and went off feed, followed by the appearance of lesions in the colon, erratic motion, oedema of the stomach and colon and muscle spasms while fish were apparently resting. A predisposition to sunburn in fish confined in the open in shallow ponds or raceways Common carp showed a congestion of the skin with subcutaneous hemorrhages. Common symptoms of niacin deficiency in most fish studied were muscular weakness and spasms, coupled with poor growth and poor food conversion.

Requirements

In homeotherms on a balanced test ration the niacin requirement is generally estimated to be about ten times that of the thiamine requirement. These rations generally contain considerable carbohydrate material to furnish energy to maintain body temperature.

In fish the requirement appears to be twenty to thirty times that of the thiamine needs determined for the same test conditions and test rations.

This difference may be due to low carbohydrate content of young fish diets and the higher protein content of these rations. Conversion of tryptophan to niacin occurs in the mammalian liver and possibly also in the liver of fish. This conversion may account for the slow development of the niacin-deficiency syndrome in fish.

However, after 10-14 weeks on diets devoid of niacin, deficiency symptoms did occur in several species of fish. The symptoms were reduced by replacement of niacin in the ration even when high protein diets containing an excess of tryptophan were fed. Too much niacin inhibits growth.

Sources and Protection

Niacin is found in most animal and plant tissues. Rich sources are yeast, liver, kidney, heart, legumes, and green vegetables. Wheat contains more niacin than corn and the vitamin is also found in milk and egg products.

The vitamin is very stable since it is generally found in coenzyme form in raw materials. Niacin added to the diet as a supplement remains relatively unaltered during diet manufacture, processing, and storage.

Antimetabolites and Inactivation

Pyridine-3-sulphonic acid and 3-acetylpyridine are compounds structurally related to niacin and are antimetabolites for this vitamin in animals and in micro-organisms. Additional niacin can overcome the anti-metabolic effect. Deficiency symptoms in rats may be induced by 6-amino niacinamide. The symptoms are reversed by addition of ten times more niacinamide than the antimetabolite. Thioacetamide has been reported to be a niacin antagonist in fish.

Clinical Assessment

Salmon feeding actively in the oceans showed liver niacin content of 70-80ωg of niacin/g of fresh tissue.

About half of this amount is present in fingerling salmon raised in fresh water at 12-15°C and fed test rations containing 40-50 per cent protein and niacin supplements 500-750 mg/kg dry diet.

Urinary metabolities of niacin have been measured in other animals on standard niacin load in test rations containing a standard tryptophan load.

The technique is well developed to measure the N1-methyl derivative in mammalian urine. These data have not been reported for fish but metabolism chambers are available for collecting branchial and urinary wastes from large fish intubated with different diet material.

BIOTIN

Certain foods, notably liver and kidney, were found to be protective against a form of dermatitis resulting from the consumption of egg white by rats, in the early thirties. It was soon learned that a specific component of egg albumin, avidin, rendered dietary biotin unavailable, hence producing the symptoms.

Biotin was variously called coenzyme R and'vitamin H'. It was isolated by Du Vigneaud in 1941 and synthesized by workers of Merck and Company in 1943.

Biotin was once thought to be part of factor H for fish. Blue slime patch disease due to biotin deficiency was reported in trout.

Biotin is a monocarboxylic acid slightly soluble in water and alcohol and insoluble in organic solvents. Salts of the acid are soluble in water. Aqueous solutions or the dry material are stable at 100 C and to light.

The vitamin is destroyed by acids and alkalis and by oxidizing agents such as peroxides or permanganate. Biocytin is a bound form of biotin isolated from yeast, plant, and animal tissues. Other bound forms of the vitamin can generally be liberated by peptic digestion. Oxybiotin has partial vitamin activity but oxybiotin sulphonic acid and other analogues are antimetabolites inhibiting the growth of bacteria.

Avidin, a protein found in raw egg white, binds biotin and makes it unavailable to fish and other animals. Heating to denature the protein makes the bound biotin available again to the fish.

Biocytin or -biotinyl lysine (the epsilon amino group of lysine and the carboxyl of biotin being combined in a peptide bond) is hydrolyzed by the enzyme biotinase making the protein-bound biotin available.

POSITIVE FUNCTIONS

Biotin is required in several specific carboxylation and decarboxylation reactions, including the carboxylation of pyruvic acid to form oxaloacetic acid. It is part of the coenzyme of several carboxylating enzymes fixing CO_2 such as propionyl coenzyme A involved in the conversion of propionic acid to succinic acid in methylmalonyl coenzyme A.

Biotin is also involved in the conversion of acetyl CO_2 to malonyl coenzyme A in the formation of long chain fatty acids. It has possible involvement in citrulline synthesis and may have effects on purine and pyrimidine synthesis. It is involved in the conversion of unsaturated fatty acids to the stable cis form in the synthesis of biologically active fatty acids.

DEFICIENCY SYNDROME

Some signs of biotin deficiency in salmonids are skin disorders, muscle atrophy, lesions in the colon, loss of appetite, and spastic convulsions. Haematology discloses fragmentation of erythrocytes. Poor growth is a common symptom and has been reported for salmonids, common carp, goldfish (Carassius auratus) and eel.

Blue slime patch disease in brook trout deficient in biotin appears typical for this species. Fish reared in 10 -15 C water exhaust biotin stores in 8-12 weeks and the first signs are anorexia, poor food conversion, and general listlessness before the more acute deficiency symptoms become detectable.

Sources and Protection

Rich sources of biotin are liver, kidney, yeast, milk products, and egg yolks. Nut meats contain good supplies of biotin. The diet should be protected from strong oxidizing agents or conditions which promote oxidation of ingredients. Raw egg white should not be incorporated into moist fish diets. Cooking will inactivate the avidin.

Antimetabolites and Inactivation

Many biotin homologues with different side chain lengths inhibit the growth of bacteria. Oxybiotin, in which sulphur is replaced by oxygen, has about the same biological activity as the natural biotin. On the other hand, oxybiotin sulphonic acid inhibits biotin activity.

Clinical Assessment

Measurement or urinary excretion of biotin in animals does not prove a good clinical method since biotin is synthesized by several organisms in the gut. In fish, this technique needs to be explored and may well be valuable as a clinical tool since the gastrointestinal tract of many freshwater fish contains only small quantities of bacteria. Biotin is one of the most expensive vitamins to add to fish rations. Salmon feeding actively in the oceans had liver biotin concentrations of 10-12 mg of wet liver tissue

. The concentrations in the liver of young salmon fingerlings in fresh water fed test diets containing an excess of biotin were between 6-8 mg of the vitamin/g of tissue. Fish with these levels of biotin in the liver should probably be in sound biotin nutritional status.

FOLIC ACID (FOLACIN)

A factor effective in curing metaloblastic anaemia in monkeys was found in yeast and liver concentrates and was designated vitamin M in 1935. In 1939 an anti-anaemic factor was found in liver and called vitamin BC. These were later shown to be the same substance with folic acid as the active ingredient. Folic acid was synthesized in 1946 and was soon used in fish diets for preventing purified diet anaemia.

Folic Acid

Folic acid crystallizes into yellow spear-shaped leaflets which are soluble in water and dilute alcohol. It can be precipitated with heavy metal salts. It is stable to heat in neutral or alkaline solution, but unstable in acid solution. It deteriorates when exposed to sunlight, or during prolonged storage.

Several analogues have biological activity including pteroic acid, rhizopterin, folinic acid, xanthopterin and several formyltetrahydropteroyl-glutamic acid derivatives. These have closely allied ring structures and many have been isolated as derivatives in various animals or microbiological preparations. One simple form, xanthopterin, present in the pigments of insects, is shown below and is of special interest because of early work with this compound as the anti-anaemic factor H for fish.

Positive Functions

Folic acid is required for normal blood cell formation and is involved as a coenzyme in one-carbon transfer mechanisms. In the presence of ascorbic acid, folic acid is transformed into the active (5-formyl-5, 6,7,8) tetrahydrofolic acid. Folic acid is involved in many one-carbon metabolism systems such as serine and glycine interconversion, methionine-homocysteine synthesis, histidine synthesis, and pyrimidine synthesis.

Several coenzyme forms of the active vitamin have been isolated. Folic acid is involved in the conversion of megaloblastic bone marrow to normoblastic type. It has a role in blood glucose regulation and improves cell membrane function and hatchability of eggs.

Deficiency Syndrome

Macrocytic normochromic anaemia occurs in several

experimental animals, including fish fed diets devoid of the folic acid. Increasing numbers of senile cells are observed as the deficiency progresses until only a few old and degenerating cells are found in the blood of deficient fish.

Anterior kidney imprints disclose only adult cells and no preforms present. Other signs observed have been poor growth, anorexia, general anaemia, lethargy, fragile fins, dark skin pigmentation, and infarction of spleen.

SOURCES AND PROTECTION

Yeast, green vegetables, liver, kidney, glandular tissue, fish tissue, and fish viscera are good sources of folic acid. Insects contain xanthopterin which has folic acid activity. At one time the yellow pigment of xanthopterin was identified as the fish anti-anaemic factor H, but subsequent experiments showed only partial activity and that folic acid itself was a much more potent antimacrocytic anaemia factor.

Insects may contribute significantly to the folic acid requirements of wild fish, but in scientific fish husbandry artificial diets are more reliable sources. Activity is lost during extended storage and when material is exposed to sunlight. Therefore, dry feeds should be carefully protected during manufacture and moist diet rations should be carefully preserved. Both types of fish diets should be fed soon after manufacture to assure minimal loss of folic acid activity.

ANTIMETABOLITES AND INACTIVATION

One antagonist of folacin is 4-aminopteroylglutamic acid or aminopterin. This material, when incorporated in the diet of guinea pigs and rats, induces anaemia and leucopenia and has been used to treat leukemia in man.

Amethopterin (4-amino-N10-methylpteroylglutamic acid) can also be used to induce deficiency by inhibiting purine, pyrimidine, and nucleic acid protection; viz., inhibition of nucleic acid synthesis results in macroytic anaemia.

CLINICAL ASSESSMENT

Haematology is used as a simple clinical tool to assess

haemopoiesis in fish. Anterior kidney imprints easily disclose normal distribution of immature cells and preforms undergoing reticulosis.

Salmon actively feeding in the ocean and young salmon fingerlings raised in fresh water and fed diets rich in folacin show liver storage of 3-4ωg of folic acid/g of wet tissue. Microbiological assay is preferred for assessment of total folic acid in dietary raw materials because the total biological activity it measures includes all the various coenzyme forms and folic acid analogues.

Assessment of the dietary intake of folic acid is important for intensive cold water fish husbandry. In pond culture, aquatic and terrestrial insects, algae, etc., may also be available and folic acid in the supplementary feed may not be as critical. Since folic acid is labile in storage, excess amounts are generally added to manufactured feed in anticipation of storage losses.

However, prudent fish husbandry dictates rapid use of manufactured rations with minimum storage. Routine periodic haematology of fish assures proper nutritional status for maximum production and sound health.

The author has noted in several series of experiments that when fish diseases occur through inadvertent contamination of the water supply, those groups of fish partially or completely deficient in folic acid were among the first to show acute disease symptoms. Therefore, folic acid must also play an important role in resistance to disease.

VITAMIN B12

The antipernicious anaemia factor found to be contained in liver was isolated and crystallized in the mid forties. This substance, named vitamin B12 by its discoverers, was later to be recognized as essential for growth of chicken fed diets entirely of plant origin and was designated animal protein factor (APF).

When anaemic salmon were injected with crystalline B12 in combination with folic acid and xanthopterin positive haemopoiesis occurred within a few days, and the salmon showed rapid recovery from the anaemia. Vitamin B12 or cynacobalamin has the following approximate chemical structure.

Cyanocobalamin

The molecule has a planar group and a nucleotide group lying nearly at right angles to one another. This cobalt-containing vitamin has a net charge of one at the central cobalt atom to which is attached a replaceable cyano group. Vitamin B12 is stable to mild heat in neutral solution, but is rapidly destroyed by heating in dilute acid or alkali. Crude concentrates are more unstable and rapidly lose activity.

The compound is similar to the porphyrins in its spatial configuration with a central cobalt atom linked to four reduced pyrrole rings in the haeme series. Replacing the cyanide ion with a variety of anions produce derivatives which have comparable biological activities; viz., hydroxocobalamin, nitritocobalamin, chlorocobalamin and sulphatocobalamin.

Positive Functions

Cyanocobalamin is involved with folic acid in haemopoiesis. It is required by many micro-organisms and is a growth factor for many animals. The animal protein factor present in fish and animal by-products was not recognized until crystalline vitamin B12 was injected into anaemic chinook salmon fingerlings in 1949 and positive haemopoiesis was observed.

A coenzyme incorporating vitamin B12 is involved in the reversible isomerization of methyl-malonyl coenzyme A to succinyl coenzyme A and in the isomerization of methylaspartate to glutomate. Cyanocobalamin is involved in the coenzyme for the methylation of homocystine to form methionine.

It is also involved in several other one-carbon reactions and in the synthesis of labile methyl compounds. One vitamin B12 containing coenzyme acts in methylation of the purine ring during thymine synthesis. Vitamin B12 is also involved in cholesterol metabolism, in purine and pyrimidine biosynthesis, and in the metabolism of glycols.

Deficiency Syndrome

Deficiency signs in young pigs, chicks, and rats show abnormal blood elements, poor growth, and pernicious anaemia. An intrinsic factor is necessary for good absorption of the vitamin from the gut.

This factor is a low molecular weight mucoprotein which normally occurs in gastric juice, and especially in hog gut mucosa. Pernicious anaemia in chinook and coho salmon is characterized by fragmented, erythrocytes with many aberrant forms present.

Haemoglobin levels are inconsistent and erythrocyte counts have a range extending from frank anaemia to a near normal blood pattern. Cyanocobalamin stores in fish tissues are slowly exhausted and only after 12-16 weeks on test do the symptoms appear in deficient salmon populations. Poor appetite, poor growth, poor food conversion, and some dark pigmentation can be observed before frank anaemia is detected.

Sources and Protection

Rich sources of vitamin B12 are found in fish meal, fish viscera, liver, kidney, glandular tissues, and slaughter house wastes. Since vitamin B12 is labile on storage, and in mild acid solution is easily destroyed by heating, care must be exercised in diet preparation containing flesh or meat scraps.

Antimetabolites and Inactivation.

The vitamin B12 coenzymes are very unstable under light, which rapidly decomposes the coenzymes. Photo-sensitivity is increased in dilute acid solutions.

Clinical Assessment

Generalized anaemia with fragmentation of erythrocytes and extremely variable haemoglobin levels and erythrocyte counts indicate possible B12 deficiency. Prompt response in individual fish is obtained by injecting B12 alone or in combination with folic acid in the ratio of 1 part vitamin B12 to 100 parts folic acid. Careful interpretation of haemotological data will enable one to distinguish one form of anaemia from the other.

ASCORBIC ACID

The isolation of ascorbic acid was due to Szent-Györgyi of Hungary and C.G. King of the U.S.A. Vitamin C synthesis was accomplished in 1933 after the chemical structure of ascorbic acid was established by British and Swiss workers. McCay and Tunison reported scoliosis in brook trout fed formalin-preserved meat in 1934 and McLaren observe haemorrhages in trout fed rations low in ascorbic acid. It was not until the sixties that a critical need for L-ascorbic acid by trout and salmon was demonstrated. L-ascorbic acid is a white, odourless, crystalline compound, soluble in water but insoluble in fat solvents.

It is readily oxidized to dehydroascorbic acid, the less biologically potent form. Ascorbic acid is very stable in acid solution because of the preservation of the lactone ring, but in alkaline solution hydrolysis occurs rapidly and vitamin activity is lost.

It is very heat labile and prone to atmospheric oxidation, especially in the presence of copper, iron, or several other metallic catalysts. The reduced form is the most biologically active but several derivatives or salts are obtainable which have varying degrees of ascorbate activity.

Positive Functions

L-ascorbic acid acts as a biological reducing agent in hydrogen transport. It is involved in many enzyme systems for hydroxylation; i.e., hydroxylation of tryptophan, tyrosine, or proline. It is involved in the detoxification of aromatic drugs and also acts in the production of adrenal steroids.

Ascorbic acid is necessary for the formation of hydroxy proline which is a constituent of collagen, a component of intercellular material in bones and soft tissues. Ascorbic acid plays a synergistic role with vitamin E as intracellular antioxidants and free radical traps. The conversion of folic acid to folinic acid requires vitamin C. Ascorbic acid is involved in the formation of chondroitin sulphate and intercellular ground substance.

Labelled ascorbic acid fed to fish previously deficient in the vitamin was shown to be rapidly mobilized and fixed in areas of rapid collagen synthesis and became concentrated in the thick collagen of the skin and in cartilagenous bones, as well as in the glands of the anterior kidney. Ascorbic acid is also involved in erythrocyte maturation.

Deficiency Syndrome

Scurvy with impaired collagen formation, perifollicular haemorrhages, loose teeth and poor bone formation, anaemia, and oedema have been reported in other animals. Deficiency signs in fish are generally related to impaired collagen formation. Fish soon show hyperplasia of jaw and snout. The same symptoms have been observed in trout, salmon, yellowtail, carp, guppies and char.

Histologically, hypertrophy of the adrenal tissues and haemorrhage at the bases of fins have been observed in coho salmon. Deficiency signs cease to develop and new growth becomes normal upon replacement of ascorbic acid in the ration. Anaemia eventually develops in extremely deficient fish and scoliosis and lordosis do not repair but are walled off by new growth around the afflicted areas of the spine when ascorbic acid is once again added to the ration.

Requirements

In the rainbow trout, reasonable blood and anterior kidney storage levels were obtained with an intake of about 100 mg of vitamin C/kg of dry ration in 10°, 12°, or 15° C water systems. When wound repair experiments were initiated, however, or when fish were exposed to other stress then the requirements doubled or tripled.

When severe abdominal or intramuscular wounds were inflicted, young fish needed at least 500 mg of active ascorbate for tissue repair comparable with control fish receiving 1 g or more of ascorbate in the diet/kg of dry diet. Coho salmon appear to need about half of these requirements for adequate tissue levels and for maximum severe wound repair rates. The requirement for ascorbic acid is related to stress, growth rate and size of the animal, as well as to the other nutrients present in the diet.

A compromise value of about 200 mg of ascorbic acid/kg diet for trout and salmon raised in freshwater systems between 10-15°C would ensure reasonable tissue storage levels and furnish some excess for mild stress conditions and for ascorbic acid loss from the diet through oxidation.

Large common carp can synthesize some ascorbate and the requirement for this species may be dependent on fish size and the environment in which they are reared.

Table. Growth and Tissue Ascorbate

Vitamin C diet treatment	Trout			Salmon		
	Average weight at 24 weeks	*Ascorbate concentrate* 1/		*Average weight at 24 weeks*	*Ascorbate concentrate* 1/	
		blood	*kidney*		*blood*	*kidney*
mg/100 g	*g*	*g/g*	*g/g*	*g*	*g/g*	*s/s*
0	2.4	2/	2/	5.0	22.3±2.2	89
5	9.6	34.4±2.9	125	6.0	30.5±1.2	132
10	10.6	34.6±1.3	137	5.7	35.8±1.6	265
20	10.1	38.8±3.3	132	6.1	34.2±2.3	183
40	10.2	46.8±6.2	162	6.3	33.7±2.0	225
100	10.8	51.0±4.6	247	6.0	37.8±2.3	321

1/ Average of five samples for blood (± S.D.) and two for head kidney tissue

2/ No fish available for assay

Sources and Protection

Ascorbic acid is widely distributed in nature with citrus fruits, cabbage, liver, and kidney tissue good sources for the vitamin. Fresh insects and fish tissues contain reasonable amounts of the vitamin. Synthetic ascorbic acid is also readily available. Fish food should be protected from oxidizing agents and kept sealed or frozen until used to prevent loss of the vitamin.

Antimetabolites and Inactivation

D-ascorbic acid, the optical isomer of the active form, has no vitamin activity and competes for sites of several enzyme reactions mediated by L-ascorbic acid. 6-deoxy-L-ascorbic acid, dehydroascorbic acid and L-glucoascorbic acid have very little activity.

Clinical Assessment

Ascorbic acid status for experimental animals is normally attempted by tissue ascorbate analysis. Most of the assays previously used measure total ascorbate and not biologically active L-ascorbic acid. Consequently these are fraught with errors and misconception of true vitamin C status.

In fish tissues, blood and liver do not adequately reflect the ascorbic acid intake and status, whereas assay of the anterior kidney which contains adrenal tissue provides a fairly representative picture of tissue storage of the vitamin. Stress rapidly reduces the ascorbic acid content of this tissue with concurrent production of adrenal steroids. Conversely, dietary repletion is reflected by up to four or five-fold storage levels from the deficient state.

Examination of fragile support cartilage in the gill filaments under low magnification will detect early hypovitaminosis before clinically acute symptoms become noticeable.

However, the best tissue for routine clinical analysis to assess vitamin C status in trout and salmon appears to be the anterior kidney with samples selected from the junction of the two forward wings. Tissues are blotted free of blood with filter paper, and then assayed for total ascorbate by one of the improved quick methods to determine total ascorbate.

Inositol

Muscle'sugar' was discovered by Scherer in 1850 and was characterized as inositol in 1887. Inositol was shown to be an alopecia (a form of hairlessness) preventing factor for mice in 1940. Poor growth and poor food passage in inositol-deficient fish was observed in salmon and carp.

Inositol

Seven optically inactive and two optically active isomers of hexahydroxycyclohexane can exist. One of the optically active forms, myo-inositol, is a white crystalline powder soluble in water and insoluble in alcohol and ether. The material can be synthesized, but is easily isolated from biological material in free or combined forms.

The mixed calcium-magnesium salt of the hexophosphate is phytin. Isomers have little biological activity but will compete in chemical reactions. Inositol is a highly stable compound.

POSITIVE FUNCTIONS

Myo-inositol is a structural component in living tissues. It has lipotropic action by preventing accumulation of cholesterol in one type of fatty liver disease and is involved with choline in maintaining normal lipid/metabolism.

It is a growth-promoting substance for micro-organisms and prevents an alopecia in mice. It is a reserve carbohydrate in muscle as well as a major component of phosphoglycerides in animal tissues.

DEFICIENCY SYNDROME

Poor growth, increased gastric emptying time, oedema, dark colour, and distended stomachs are symptoms observed in salmon, trout, carp and catfish held for long periods on inositol-deficient test rations.

A'spectacle-eye' condition for rats has not been observed under the experimental conditions used in fish studies. The major deficiency sign is inefficiency in digestion and food utilization and concomitant poor growth leading to a population of fish with distended abdomens.

SOURCES AND PROTECTION

Myo-inositol occurs ubiquitously in large amounts wherever biological tissue is found. Wheat germ, dried peas, and beans are rich sources.

Brain, heart, and glandular tissues are very good sources

of biologically active inositol. Citrus fruit pulp and dried yeast also contain inositol. The compound is stable.

ANTIMETABOLITES AND INACTIVATION

Seven optically inactive and one optically active but biologically inactive stereo isomers occur. Since inositol is synthesized in the biologically active form by many microorganisms in the gut, only large quantities of chemically synthesized inactive isomers added to diets would interfere with the metabolism of inositol for growth and normal physiological function.

Biologically inactive stereoisomers of myo-inositol do not compete for critical sites in metabolism. Methyl derivatives and mono-, di- and triphosphoric acid esters occur naturally. Salts of the hexaphosphate or phytin make the bound inositol practically unavailable to the animal.

CLINICAL ASSESSMENT

Assessment in fish has been based on lack of deficiency signs coupled with the most efficient food conversion. Salmon feeding actively in oceans show 1-1.5 mg of inositol/g of fresh liver tissue and young fingerlings raised in freshwater at 10-15°C had 600-700ωg/g of liver tissue.

An alternate, better assessment may be based on a standard muscle section or whole carcass analysis for free or for bound inositol. Projection of inositol intake from normal fish diet ingredients should indicate an excess of this particular vitamin.

CHOLINE

Methylation as a basic metabolic process was postulated by Hoffmeister. Methyl transfer was shown in vivo by Thompson and the interrelationships between choline, methionine, and homo-cystine were shown by du Vigneaud in 1939-42.

Trout fed low choline rations developed haemorrhagic kidneys and salmon showed an aversion to food in choline-deficient diet.

Choline

Choline is a very strong organic base and forms many derivatives widely distributed in animal and vegetable tissue. One derivative, acetylcholine, is involved in the transmission of nerve impulses across synapses. Choline is very hygroscopic, very soluble in water, and is stable to heat in acid, but decomposes in alkaline solutions.

Positive Functions

Choline acts as a methyl donor in trans-methylation reactions. It is a lipotropic and antihaemorrhagic factor preventing the development of fatty livers.

It is involved in the synthesis of phospholipids and in fat transport. Acetylcholine transmits the excitory state across the ganglionic synapses and neuromuscular junctions. Choline is essential for growth and good food conversion in fish.

Deficiency Syndrome

Deficiency signs include poor growth and poor food conversion accompanied by impaired fat metabolism. Haemorrhagic kidneys and intestines have been reported in trout and increased gastric emptying time has been observed in salmon.

Sources and Protection

Rich sources of choline are wheat germ, beans, brain, and heart tissue. Choline hydrochloride, the commercially available form, may inactivate -tocopherol and vitamin K when in direct contact with these vitamins and care should be exercised in selecting properly protected (gelatin coated) fat soluble vitamins in fish diet preparation.

Clinical Assessment

Choline status of fish can be estimated from the assay of choline content of the dietary ingredients with the absence of deficiency signs. Maximum liver storage may not be the best criteria to determine choline nutritional status but has been used to assess the tentative requirement listed for the two species of salmon.

p-AMINOBENZOIC ACID

Para-aminobenzoic acid is a white crystalline powder which is water soluble and heat and light stable in aqueous and mild alkaline solution. It is a growth promoting vitamin for micro-organisms which require the vitamin for folic acid synthesis. Large intake has been shown to counteract the antimetabolite effect of sulphonamides in bacterial culture. No positive function or deficiency signs have been observed in fish.

LIPOIC ACID

Lipoic acid is both fat soluble and water soluble. Its active form is the amide derivative known as lipoamide. Lipoic acid functions as a coenzyme in a -ketoacid decarboxylation. It was discovered independently in several laboratories during the period 1945-50 and shown to be an essential component of multienzyme, the pyruvate dehydrogenase complex. Pyruvate is converted to'active acetaldehyde' which in turn is picked up by lipoamide.

Subsequent oxidation of the aldehyde results in the reduction of lipoamide to a disulphydryl form. The multi-enzyme unit also includes thiamine pyrophosphate, coenzyme A, and flavin adenine dinucleotide. Glandular tissues are good sources of lipoic acid. No requirements have been determined for fish.

FAT-SOLUBLE VITAMINS

Fat-soluble vitamins A, D, E, and K differ from the water-soluble vitamins in their accumulative action. Little evidence has been recorded for hypervitaminosis with the water-soluble vitamins since these compounds are rapidly metabolized and excreted when intake exceeds liver or tissue storage capacity, but hypervitaminosis is a common occurrence in fish and other animals when large quantities of any one of the fat-soluble vitamins are ingested.

Toxicity symptoms involving vitamins A and D are indistinguishable from deficiency symptoms for the same vitamins. On the other hand, symptoms of excess vitamins E and K intake are more discrete. Fish rations may often be enriched with fish oils to increase caloric density of the ration resulting in excessive intake of the fat-soluble vitamins.

VITAMIN A

A fat-soluble factor that promoted growth in rats by Hopkins and by Osborne and Mendel at the beginning of this century. McCollum and Simmonds cured xerophthalmia, a disease of the eye, with this material. The chemical structure of vitamin A and its relationship to -carotene was shown by Von Euler in 1928. Active vitamin A was synthesized in the mid-thirties.

The relationship of the vitamin A alcohols to naturally occurring -carotene Retinene, the aldehyde form of vitamin A, has been isolated from the retina of dark adapted eyes and is involved in vision in dim light. Retinoic acid, which is the oxidized form of vitamin A alcohol, has been shown to have some vitamin A activity. Vitamin A1 is found in saltwater fish, whereas vitamin A is more abundant in freshwater fish. Interconversion of the two forms in living fish tissue has been demonstrated.

Fish oils contain vitamin A as free alcohols or esters. Vitamin A alcohol occurs as a light coloured viscous oil which is heat labile and subject to air oxidation. Beta-carotene occurs as an orange, crystalline compound which is more stable to heat and oxidation. Vitamin A is water insoluble but is soluble in fat and organic solvent.

POSITIVE FUNCTIONS

Vitamin A is essential in maintaining epithelial cells, preventing atrophy and keratinization of epithelial tissues. It is combined with a protein in visual purple and is important in night vision. Vitamin A also prevents xerophthalmia in rats and young children. Vitamin A promotes growth of new cells and aids in maintaining resistance to infection.

THE SYNDROME OF DEFICIENCY OR EXCESS

Hypovitaminosis A is characterized by poor growth, poor vision, keratinization of epithelial tissue, xerophthalmia, night blindness, haemorrhage in the anterior chamber of the eye, haemorrhage of the base of the fins, and abnormal bone formation. Nerve degeneration has been reported in pigs, chickens, rats, rabbits, and ducks, but only occasionally observed in fish after long periods of deficiency.

Hypervitaminosis A fish and in other animals and involved enlargement of liver and spleen, abnormal growth, skin lesions, epithelial keratinization, hyperplasia. of head cartilage, and abnormal bone formation resulting in ankylosis and fusion of vertebrae.

Hypervitaminosis A is reflected in very high liver oil vitamin A content and elevated serum alkaline phosphatase. Removal of excess vitamin A from the diet promotes rapid recovery.

SOURCES AND PROTECTION

Cod liver oil is the best known source of vitamin A although black sea bass, swordfish, or ling cod oils contain much more of the vitamin. Whale liver oil contains kitol which has little or no biological activity until heated above 200° C when one molecule of biologically active vitamin A is generated per molecule of whale kitol.

This biologically inactive kitol may be deposited in the whale to avoid hypervitaminosis A during excessive vitamin A intake. The possibility of hypervitaminosis A occurs when tuna, shark, or ling cod viscera are used in preparation of moist fish diets. Synthetic vitamin A preparations, such as vitamin A palmitate, are available and are often used to supplement rations low in fish meal, fish viscera, or carotenes.

Some fish species seem able to utilize b-carotene as a vitamin A source, whereas others are unable to split the g-carotene molecule and vitamin A must be added to the diet.

CLINICAL ASSESSMENT

Vitamin A status can best be assessed by the absence of deficiency signs and by assay of liver oil for vitamin A content. Assay for the vitamin in blood or plasma has not been found to be useful.

VITAMIN D

Rickets was induced with test diets by Hopkins, and Mellanby cured the disease in dogs by adding cod liver oil to the ration. Ultraviolet light exposure was shown to have an anti-rachitic effect, and provitamin D activity had been ascribed to ergosterol. Crystalline vitamin D was isolated by Angus and activated 7-dehydrocholesterol was isolated by Windaus.

Vitamin D_2 or ergocalciferol, is one of several biologically active forms of vitamin D. Vitamin D_3 or activated 7-dehydrocholesterol has the chemical formula $C_{27}H_{44}O$ and contains a more simplified, unsaturated 8-carbon side chain. Vitamin D_3, also known as cholecalciferol, is formed in most animal tissue by the rupture of one of the ring bonds of 7-dehydrocholesterol by ultraviolet radiation.

Cholecalciferol is a white, crystalline compound soluble in fat and organic solvents and is stable to heat and oxidation in mild alkali or acid solution.

POSITIVE FUNCTIONS

Vitamin D is essential for maintaining calcium and inorganic phosphate homeostasis. It is involved in alkaline phosphatase activity, promotes intestinal absorption of calcium, and influences the action of parathyroid hormone on bone.

Fish may sequester calcium from water through the gill membrane; thus the major function of vitamin D for other animals may not be necessary to satisfy calcium requirements for fish.

SYNDROME OF DEFICIENCY OR EXCESS

Rickets and abnormal bone formation for animals has not been observed in fish fed low vitamin D diets. However little work has been done under carefully controlled conditions with young growing fish and only alteration in alkaline phosphatase activity has been reported on different vitamin D diet intake.

Hypervitaminosis D, however, has been reported. Brook trout fed large doses of vitamin D showed impaired growth, lethargy, and dark colouration. High intake of vitamin D mobilizes phosphorus and calcium from the bone and tissues and may result in fragile bones, poor growth, and poor appetite related to the nausea in man afflicted with hypervitaminosis D.

This area needs to be explored because of the potential for hypervitaminosis D in fish fed diets containing various fish viscera which might contain large amounts of the vitamin D. Tuna liver oil may contain, for example, 100 to 1 000 times as much active vitamin D as cod liver oil.

SOURCES AND PROTECTION

Vitamin D requirements of many animals can be met by ultraviolet irradiation of the skin where cholesterol derivatives are converted to calciferol. Since the vitamin is fat soluble and accumulates in lipid stores, fish liver oil is a rich source of the material. Content varies tremendously in liver oil, however, with values of about 25 I.U./g present in shark liver oil and over 200 000 I.U./g in albacore tuna liver oil. Cod liver oil contains from 100-500 I.U./g and animal liver contains some vitamin D. One international unit (I.U.) is equal to 0.025 mg of crystalline vitamin D.

CLINICAL ASSESSMENT

Absorption maxima in the ultraviolet region can be used to detect provitamins D in the non-saponifiable fraction of oils. Concentrated preparations of vitamin D can be assayed by the Carr-Price antimony trichloride reaction when assay is necessary to determine biologically active materials in liver oil of fish on different treatments.

The chick assay may not apply to fish liver storage levels because the biologically active form for fish has not been determined. Therefore, clinical assessment for hypervitaminosis A must rely on crude methods for determination of vitamin D in liver oil samples by the Carr-Price reaction or by measuring absorption in the UV spectrum.

VITAMIN E

The existence of an antisterility vitamin was postulated by Evans and Bishop. This factor was named'vitamin E' by

Sure. The active tocopherol was isolated, characterized, and synthesized by Karrer. Vitamin E is composed of a class of compounds known as tocopherols. One of the most important tocopherols is -tocopherol.

Eight naturally occurring tocopherol derivatives have been isolated and all belong to the D series. Synthetic -tocopherol is a racemic DL- -tocoperhol mixture. The derivatives of tocol or of tocotrienol are named, alpha, beta, gamma, delta, epsilon, eta, zeta1 and zeta2-tocopherol.

The pure tocopherols are fat-soluble oils which are capable of esterification to form crystalline compounds. The tocopherols are stable to heat and acids in the absence of oxygen, but are rapidly oxidized in the presence of nascent oxygen, peroxides, or other oxidizing agents.

The tocopherols are sensitive to ultraviolet light and are excellent antioxidants in the free form. The esters are more stable and are commonly used as dietary supplements-anticipating hydrolysis in the gut and absorption of the free alcohol to act as an active intra- and intercellular antioxidant. Ethyl derivatives on the aromatic ring are also active.

Oxidation products of a-tocopherol can be reduced with hydrosulphite to -tocopherylhydroquinone or, in the presence of ascorbic acid, to -tocopherol.

Table. The Tocopherols

Tocopherol (alpha)	5,7,8-trimethyltocol
Tocopherol (beta)	5,8-dimethyltocol
Tocopherol (gamma)	7,8-dimethyltocol
Tocopherol (zeta$_2$)	5,7-dimethyltocol
Tocopherol (eta)	7-methyltocol
Tocopherol (delta)	8-methyltocol
Tocopherol (epsilon)	5,8-dimethyltocotrienol
Tocopherol (zeta$_1$)	5,7,8-trimethyltocotrienol

POSITIVE FUNCTIONS

The tocopherols act as extra and intracellular antioxidants to maintain homeostasis of labile metabolites in the cell and tissue plasma. As physiological antioxidants, these usually protect oxidizable vitamins and labile unsaturated fatty acids. Vitamin E prevents encephalomalacia in chicks, erythrocyte haemolysis in several animals and steatitis in mink, pigs, and farm animals.

The tocopherols also prevent exudative diathesis in chicks, white muscle disease in lambs and calves, and dietary liver necrosis in rats. They act with selenium and with vitamin C for normal reproductive activity and are involved in prevention of nutritional muscular dystrophy in the chick, the yellowtail and carp.

The tocopherols act as free radical traps to stop the chain reaction during peroxide formation and stabilize unsaturated carbon bonds of polyunsaturated fatty acids and other long-chain labile compounds.

Vitamin E is involved in the maintenance of normal blood capillary permeability and the integrity of heart muscle. It was first shown to be involved in prevention of sterility and foetal resorption in rats. It may likewise affect embryo membrane permeability and hatch-ability of fish eggs.

SYNDROME OF DEFICIENCY OR EXCESS

One of the first signs is erythrocyte fragility closely followed by anaemia, ascites, xerophthalmia, poor growth, poor food conversion, epicarditis, and ceroid deposits in spleen and liver. Muscle dystrophy and xerophthalmia have been in yellowtail and carp. Non-specific forms of degenerative conditions have been described in several species of fish fed large quantities of polyunsaturated fatty acids with inadequate tocopherol in the ration. Hypervitaminosis E results in poor growth, toxic liver reaction, and death.

SOURCES AND PROTECTION

Vegetable oils are rich sources of tocopherols. Synthetic - tocopherol in esterfied acetate or phosphate form is commonly used as a diet supplement. These esters are much more stable than the free form which is rapidly lost by air oxidation or in the presence of labile compounds like the polyunsaturated fish oils.

Addition of antioxidants such as BHT (butylated hydroxyanisole) and BHT (butylated hydroxytoluene) protects fats and other labile compounds in the ration from oxidation, but these antioxidants have no vitamin E activity.

CLINICAL ASSESSMENT

The erythrocyte fragility test indicates the physiological state of fish. Absence of histologically detectable ceroid in liver and spleen from representative samples in the population is a good clue on the presence of adequate amounts of physiological antioxidants in the fish.

A barbituric acid test for oxidation of components has not been applied to fish tissues as a clinical tool for assessment of nutrition state except when liver oils become saturated with very labile polyunsaturated fatty acids such as when feeding squid or saury oils.

These assays are good indicators of the state of oxidation of the finished ration, but the absence of deficiency signs and normal erythrocyte fragility are better clinical tests. Analysis for tocopherol is difficult and time-consuming and only applicable under critical research experiment situation.

VITAMIN K

The name vitamin K (for'koagulation') was proposed by Dam. Dam isolated the vitamin from alfalfa and from fish meal in 1939. It was synthesized later that year. The K vitamins K_2 contain 6,7, or 9 isoprene units in the side chain which varies from 30-45 carbon atoms. Many isosteres of vitamin K have been identified in animal tissues, plant tissues, and micro-organisms.

Although fairly stable compounds, they are destroyed upon oxidation and exposure to ultraviolet radition. Menadione is very reactive and is subject in aqueous media to chemical interaction.

POSITIVE FUNCTIONS

Vitamin K is involved in the hepatic synthesis of blood clotting proteins - prothrombin and proconvertin. Substituted forms of vitamin K are strongly bacteriostatic and may serve as an alternate defence mechanism for bacterial infections.

Vitamin K is structurally related to flavo compounds which function as electron carriers in the electron transport system. The primary role of vitamin K is to maintain a fast normal blood clotting rate which is so important to fish living in a water environment.

SYNDROME OF DEFICIENCY OR EXCESS

Prothrombin time in salmon fed diets devoid of vitamin K was increased three to five times and, during prolonged deficiency states, anaemia and haemorrhagic areas appeared in the gills, eyes, and vascular tissues.

Increased blood clotting time has also been reported for other fish reared on diets with low vitamin K content. Interrelationships with other vitamins have not been documented in fish where the primary deficiency signs are slow blood clotting and haemorrhage, severe anaemia, or death in wounded fish. Haemorrhagic areas often appear in fragile tissues such as the gills.

SOURCES AND PROTECTION

Vitamin K sources are green, leafy vegetables. Alfalfa leaves are among the best sources of vitamin K. Low levels are found in soybeans and animal liver. Synthetic menadione is also available.

Although vitamin K in ground alfalfa is fairly stable, synthetic material should be protected from exposure to ultraviolet light and to excessive oxidizing or reducing conditions. The use of rapidly cured dehydrated alfalfa is essential to minimize formation of the physiological antagonist, dicumarol.

The diet should be kept dry, prepared with minimum exposure to air oxidation, and fed as soon as practicable after manufacture to minimize vitamin K loss.

ANTIMETABOLITES AND INACTIVATION

Dicumarol was isolated from spoiled sweet clover hay and

shown to prevent the normal function of vitamin K in maintenance of normal blood clotting times. Early work was with cattle and hogs but was soon extended to rats, other experimental animals, and man.

Dicumarol is an anticoagulant and has been used to prevent thrombosis in animals and man. It is not an antimetabolite competing for vitamin K sites but plays another role in preventing normal blood clotting.

Vitamin K counte-racts the dicumarol effect quantitatively. Another anticoa-gulant, warfarin, is a common rat poison which has five to ten times the anticoagulant activity of dicumarol. The effects of warfarin can also be reversed by administration of vitamin K.

CLINICAL ASSESSMENT

Vitamin K status of fish is determined by measuring blood clotting time. Chemical methods for determining menadione content of active metabolite after storage are available but have not been applied to measure vitamin K activity in the liver or tissues of young, growing fish.

Chapter 5

Chemistry of Sugars

Sugars-or saccharides- are the most abundant biomolecule on the planet. They are important in a number of biological roles. Most obviously you will know them as a major component of your diet.

Insoluble sugars also function as structural material in the cell walls of plants and bacteria and in the connective tissue and cell coats of animals.

Sugar poiymers serve to lubricate skeletal joints. Sugars can be attached to proteins (in glycoproteins) and lipids (in glycolipids) and these *glycosylated* compounds serve:

- As antigenic sites;
- Provide signals that determine the cellular localization of proteins
- Function as signals that allow cells to recognize each other and adhere in the formation of tissues and organs.

You will already have encountered two sugars as structural components of nucleic acids; these are ribose and deoxyribose, A simple sugar is a derivative of a straight chain polyhydroxy-alcohol.

Two classes of derivatives exist. In the first class, a terminal (primary) alcohol is oxidized to an aldehyde to yield a member of the ALDOSE series.

In the second class, a penultimate (secondary) alcohol function is oxidized to a carbonyl to yield a member of the KETOSE family. We can consider glycerol to be the parent of all sugars.

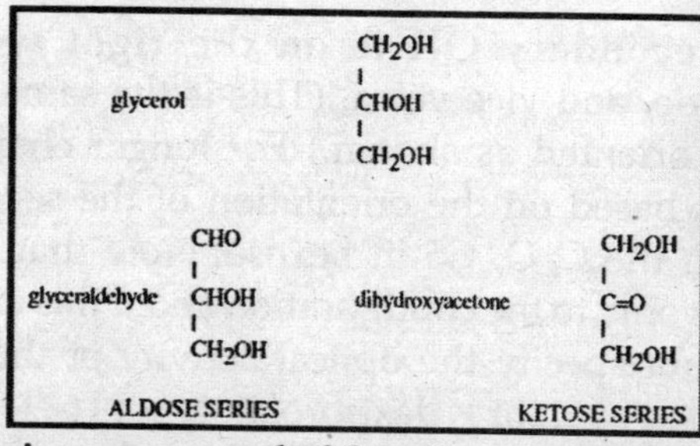

Both series are extended by the consecutive addition of CHOH functions. This is most simply viewed as starting from the longer homologs of glycerol and each time oxidizing the terminal (aldoses) or penultimate (ketoses) alcohol function.

For example:

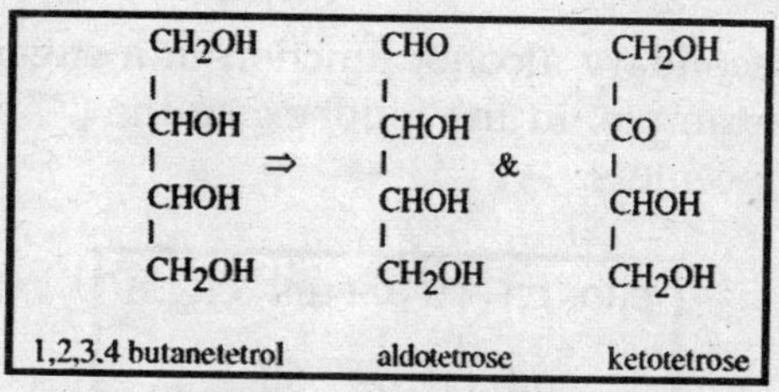

Nomenclature: Prefix: aldo- (-CHO); keto- (-C=O) Suffix: -ose (=sugar). No. of C atoms = tri (3), tetr (4), pent (5), hex (6). E.g. aldohexose = 6 C sugar in CHO form. Sugars are numbered starting from the end that has been oxidized.

In glyceraldehyde, the middle carbon atom is chiral; i.e., it bears four different substituents, and consequently glyceraaldehyde has non-superimposable stereo isomers; it possesses enantiomeric (mirror-image) forms.

The D- and Lforms of glyceraldehyde are identified as follows. Orient the glyceraldehyde as shown below with the eye of the observer viewing from the primary alcohol towards the aldehyde.

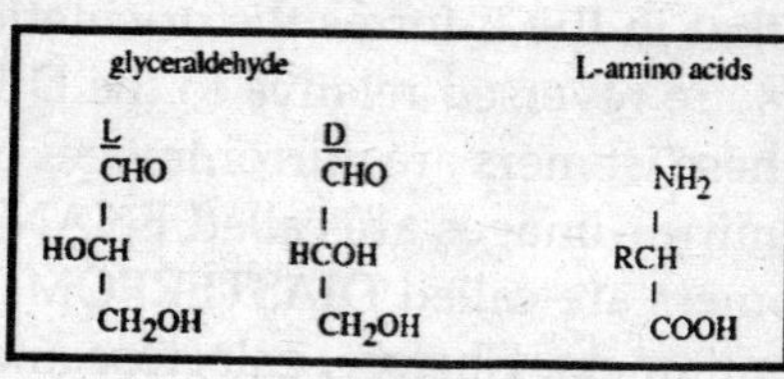

If the secondary OH is on the right we have D-glyceraldehyde, and vice versa. (This is the same rule as for amino acids oriented as shown.) For longer chains the D/L distinction is based on the orientation of the secondary OH furthest from the C=O, C5 in hexose. Note that D/L in this context refers only to the configuration about this carbon atom.

It does not specify the optical activity of the sugar; the latter is denoted by d (or +, dextrorotatory) or l (-, levorotatory). The D/L system has now been officially replaced by the Cahn-Ingold-Prelog or R, S system for designating configuration at sp^3 carbon centers (this is reviewed in Additional Material). D-sugars = R-sugars.

STEREOCHEMICAL ISOMERS

Every secondary alcohol function in a sugar is a chiral centre. For example, in the aldohexose the "*" identify the chiral carbon centers.

```
CHO - CH - CH - CH - CH - CH2OH
      |    |    |    |
      OH   OH   OH   OH
```

N chiral centers yield 2N isomers; there are consequently 16 isomers of aldohexose. These are divided into 2 families, D/L, depending on the configuration at C5 (the last starred carbon on the right). The D-isomers are the important ones. We will deal mainly with one of them, D-GLUCOSE, which is the most abundant monosaccharide from the aldose family, but two others, D-GALACTOSE (a component of milk sugar) and DMANNOSE, are important biochemically.

These sugars are shown below using a shorthand form of the Fischer projection formula; the sideways-pointing lines identify the orientation of the OH substituent of the secondary alcohols. Note that in the L-forms the orientations of all four secondary OH's are reversed relative to the D form.

Not all of these isomers are mirror images of one another. Those that are mirror-images are called ENANTIOMERS; the other stereo-isomers are called DIASTEREOMERS. D- and L-glucose are enantiomers. Glucose, galactose and mannose are diastereomers. D-Glucose and D-mannose, which differ by the orientation of a -OH at a single chiral centre, are called EPIMERS (epimers are thus a subset of diastereomers). Mannose and galactose are epimers of glucose but not of each other.

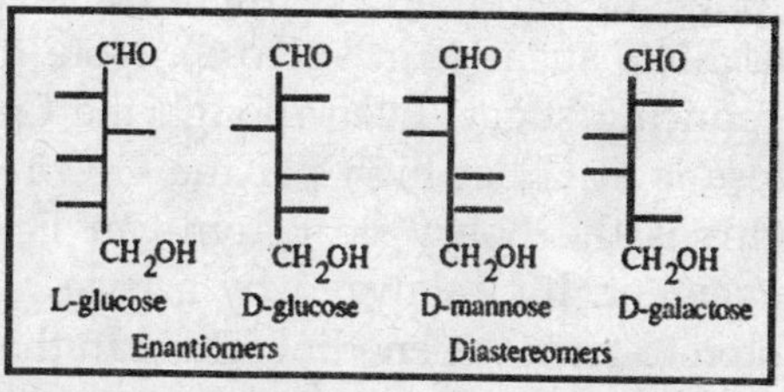

Other sugars you will (or already have) encounter are Threose (4C), Ribose and Deoxyribose (5C), Fructose (6C ketose) and Sedoheptulose (7C ketose). As ketoses have one less chiral centre than hexoses they have half as many structural isomers for the same chain length; otherwise fructose resembles glucose with the C=O moved down 1 position.

Chemical Reactions Important for Structural Analysis

- Ether formation: Base-induced reaction of all -OH's with alkylating agents e.g., methyl iodide plus dimethyl sulfoxide (anion, scavenges proton) plus sodium hydride (catalyst; very strong base) to produce methyl ether (-OCH_3; -OH's react as strongly nucleophilic -O-). The ether link is very stable. The oxygen at C1 also reacts (it is a hemi acetal which will be clarified soon) but the -OCH_3 at C1 is atypical and hydrolyzes to -OH in dilute acid. Because of this the product from reaction of glucose is 2,3,4,5,6 pentamethyl glucose.
- Reduction (with borohydride) converts CHO to CH_2OH (aldoses and ketoses give an alditol) e.g. glucose to glucitol (a.k.a. sorbitol); fructose gives glucitol *and* mannitol.
- Oxidation. Mild oxidation is often used to test for the free C=O group because with weak oxidizing reagents such as alkaline Ag^+ (Tollen's Reagent) or

alkaline Cu^{2+} (Fehling's Reagent) the products are the metallic Ag or the red insoluble Cu_2O both of which can be easily visualized. Sugars that give a positive response in these tests are called REDUCING sugars. If the -C=O group is tied up so that it cannot isomerize to a free -C=O then we have a NON-REDUCING sugar (e.g. sucrose). Note that fructose is a reducing sugar. Even though the C=O group of fructose is at C_2, at high pH the C=O rearranges to C1; this is the Lobry de Bruyn-van Eckenstein re-arrangement. It's catalyzed by a base removing the a-proton to form the enediol. This can then rearrange to yield fructose or re-capture the proton to yield glucose + mannose (because the return of the proton can be to either face).

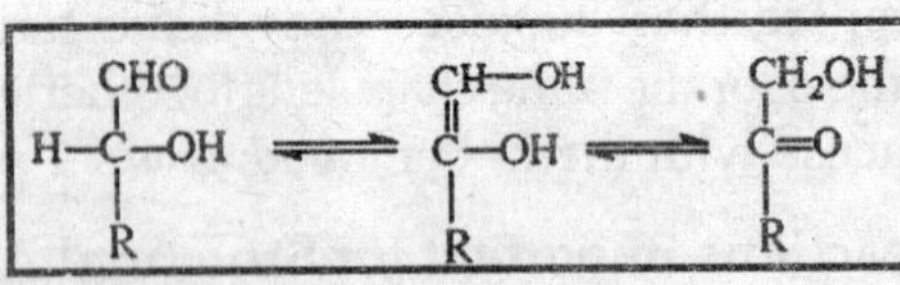

This re-arrangement is found twice in the glycolytic pathway (where it is enzyme catalyzed).

Some nomenclature:

- CHO to COOH (aldose to aldonic acid) e.g., gluconic acid
- CH_2OH to COOH (aldose to uronic acid) e.g., glucuronic acid.
- CH_2OH and CHO to COOH (aldose to aldaric acid) e.g. glucaric acid.

Sugars are Rings in Solution

Up to this point we have been representing the sugars as "linear" molecules. In reality the linear form is normally a minor species (often less than 0.1%) The initial evidence for this came from the following anomalous chemical behaviour. The typical reaction of carbonyl-containing compounds with weak methylating reagents, e.g. MeOH/HCl, yields a dimethyl derivative called an acetal.

```
                reversible hydration
                                    OH                              OMe
R-CH=O ---------------------------> RCH -----MeOH/HCl---> RCH
                dehydration         OH                              OMe
                              aldehyde hydrate                    acetal
```

But with glucose, we obtain two different monomethyl products. This result is evidence that the C=O is abnormal. In fact, we know that it is "tied-up" as a *cyclic* hemiacetal. To visualize this we recognize that the aldehyde group is planar (sp^2 hybridization) and we can therefore think of two extreme orientations shown below:

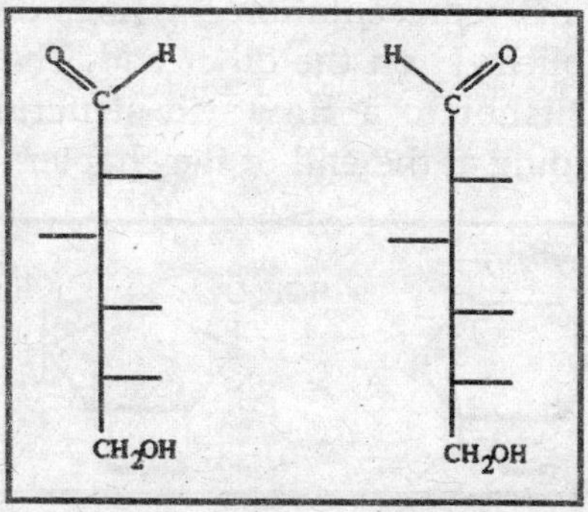

Ring closure occurs by attack of a secondary -OH (presumably as the nucleophilic anion) on the carbon of the electron deficient C=O. This attack can occur on either face of the planar CHO with the result that the -OH group that is created at C_1 can be oriented in either of two directions (if attack is on the left structure the -OH created will point to the left and vice versa). The two forms of glucose that are formed are called anomers and the C bearing the C=O is the anomeric carbon. When the newly created -OH has the same orientation as the -OH that did the attacking (the two -OH's are cis) we have the a-anomer, otherwise it is the b-anomer.

Usually the -OH that does the attacking is located on C_5 (indicated with the *) and a 6-membered ring is formed. This is called a pyranoside by analogy to tetrahydopyran. Note that by closing the ring we have created another chiral centre so that the ring forms of aldohexoses have twice as many isomers as the straight chain forms.

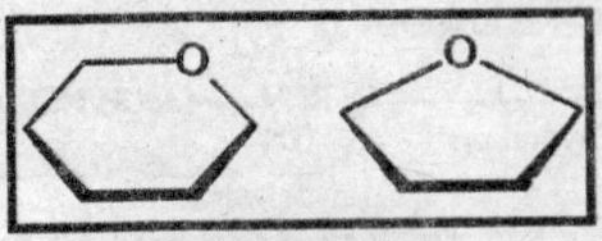

Less common is attack by the C_4 -OH which leads to the sterically strained, 5-membered furanoside (by analogy to tetrahydrofuran). In ketohexoses the C=O is at C_2 and so attack by C_5 still yields a 5-membered ring (as you will have seen in ribose).

The cyclic sugars are normally represented using Haworth structures; in this representation the lines denote -OH and there is an implicit -H on the other end. The procedure for converting a Fisher to a Haworth structure is given in Additional Reading at the end of the chapter.

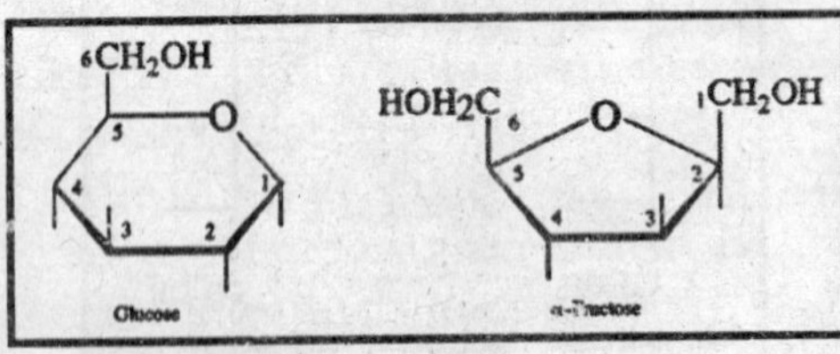

An important result of the formation of the cyclic hemiacetal is that the hydroxyl group "created" on the carbon atom, C_1, bearing the original aldehyde function can be oriented in either of two ways, i.e., a new chiral centre is created.

Thus the a and b forms just described are geometric isomers of each other. a and b glucose can be crystallized separately and can be distinguished by a measurement of optical activity. For glucose, both a and b forms are dextrorotatory.

However, dissolving the pure (a or b) solid in water leads to the production of the mixture, a phenomenon called mutarotation because the optical activity changes from that of the pure form observed immediately on dissolution to that of the equilibrium mixture some five hours later.

$$\alpha - \text{glucopyranoside} \Leftrightarrow \text{Open from} \Leftrightarrow \beta - \text{glucopyranoside}$$

40% <.1% 60%

Pure a = +113°, pure b = +19°, equilibrium solution = +52°. The ring forms of the sugars are the ones that predominate in solution. For example, the proportion of the straight-chain form of glucose is quite small, about 0.1% of the total molecules.

Glucose slowly inter converts between the three structures (half time ~ 1 hour). Note however that when a sugar is bound to the active site of an enzyme it is frequently converted to the straight-chain form.

Di-, Oligo- and Poly-saccharides

An extremely important biochemical reaction is the condensation of two (or more) monosaccharides by the elimination of water from an OH group present on each of the two sugars.

Most commonly the reaction occurs between the OH present on C_1 of one monosaccharide and that present on C_4 of the second to form a $1 \Rightarrow 4$ GLYCOSIDIC linkage. Because the reaction involves C_1, which can exist in either α- or β-forms, we can obtain either an α (1⇒ 4) or a β (1⇒ 4) glycoside.

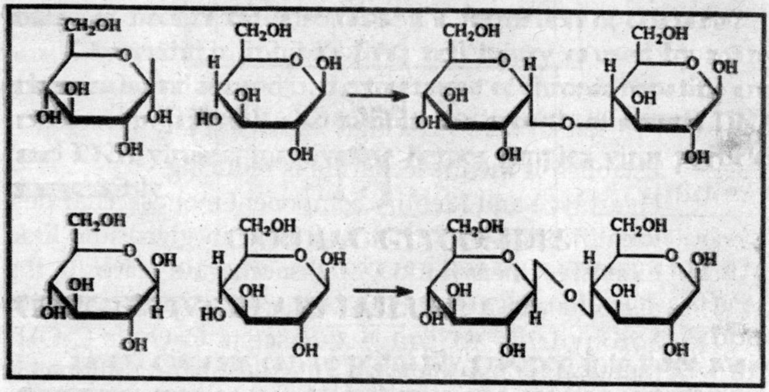

Important Disaccharides: The dimer of glucose with the Fig α(1 ⇒ 4) glycosidic link is the disaccharide MALTOSE. It is an intermediate in the hydrolysis of STARCH; its full name is α-D(glucopyranosyl) α(1⇒ 4)-D-glucopyranose (top right, also next left).

This reminds us that the "right-hand" sugar will exist in

both α and β anomeric forms. The picture shown is the β-anomer. The dimer of glucose with the Fig β(1⇒ 4) glycosidic link is called cellobiose; i t is an intermediate in the degradation of cellulose (grass and tree trunks).

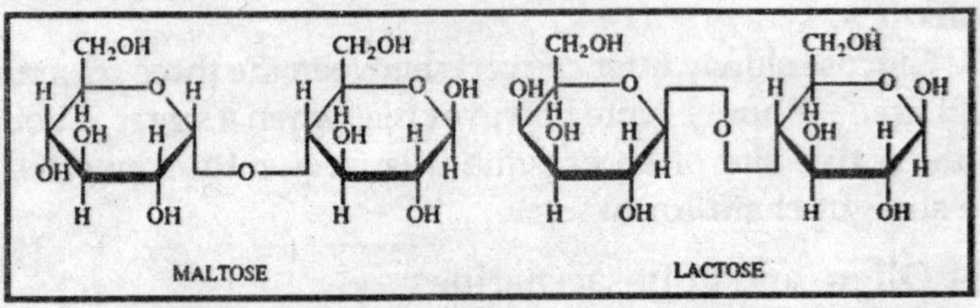

The disaccharide LACTOSE present in milk is galactose Fig β(1⇒ 4) glucose with the full systematic name of β-D (galactopyranosyl)- (1 4)-D glucopyranose. Glycoside links to other carbon atoms are fairly common, notably 1α(1⇒ 2 and 1 ⇒ 6. SUCROSE (domestic sugar) is glucose Fig (1 ⇒ 2) -β-fructose.

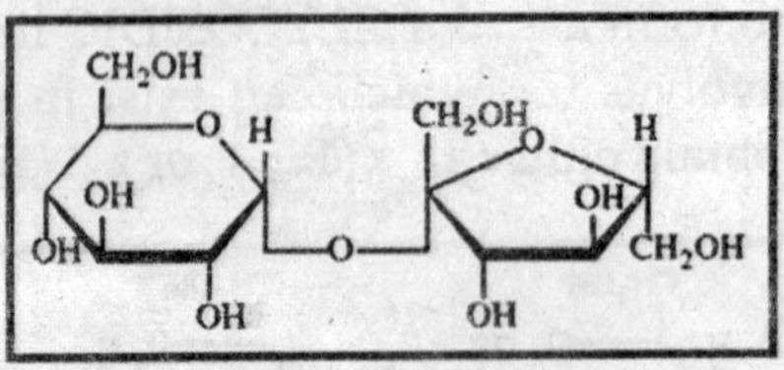

Structure Determination of Disaccharides

- Establish if the disaccharide is reducing.
- Hydrolyze and identify component monosaccharides.
- Identification of C atoms present in glycosidic link. Extensive methylation of disaccharide (care: is the disaccharide a reducing sugar? If so pre-treat with borohydride to convert nascent C=O to C-OH) followed by hydrolysis of glycosidic bond (mild acid) and subsequent identification of the methylated monosaccharides (gas liquid chromatography + mass spectrometrys).
- Nature of glycosidic link. Which enzyme hydrolyzes the disaccharide?

  Yeast Maltase Þ a (This one)

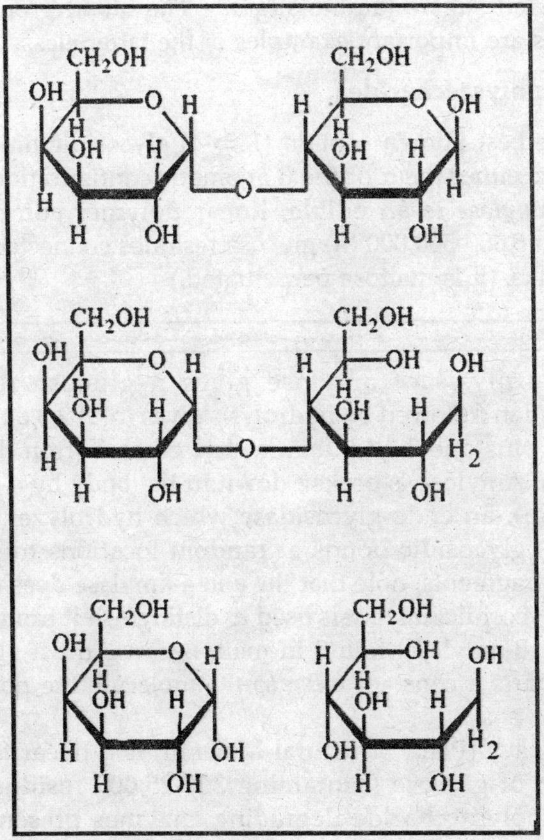

Almond Emulsin Þ b (Not this one)

POLYSACCHARIDES

By the consecutive application of the glycosidic condensation reaction we obtain polysaccharides. In contrast to proteins and nucleic acids which are linear polymers with variable monomeric units, the multiple -OH groups present on the sugar ring allow the possibility of both linear and branched polymers most commonly of the same monomeric unit, namely glucose.

When only a single monosaccharide is present the polysaccharide is called simple: (e.g., starch and cellulose). Complex polysaccharides contain several sugars, including substituted sugars (amino-sugars). The blood group (ABO) antigens are important examples of the latter class.

Linear polysaccharides.

The best known contain (1 ⇒ 4) glycosidic links. These can have either the α or the β anomeric configuration but not both. *Amylose* is an edible, linear polymer composed of between 5,000-500,000 (n) glucose residues connected by α (1 ⇒ 4) links. (it is maltose perpetuated.)

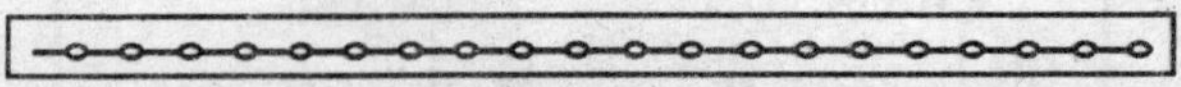

Hydrolysis of amylose gives n glucose residues; methylation followed by hydrolysis gives (n-1) 2,3,6 trimethyl glucose plus one 2,3,4,6 tetramethyl glucose (from left-most residue). Amylose is broken down in the body by α-amylase (in saliva), an endo-glycosidase which hydrolyzes internal α(1⇒ 4) glycosidic bonds at random locations to produce smaller fragments; note that the a in a-amylase does not refer to the glycosidic link but is used to distinguish it from another enzyme, β-amylase, found in malt. Because the a glycosidic link imparts a constant twist to the molecule the polymer is helical.

Cellulose (Plant Structural Material) is a linear β(1 ⇒ 4) polymer of glucose (containing 300-25,000 residues). The human polysaccharide degrading enzymes present in the saliva and pancreatic juice hydrolyze α(1 ⇒ 4) but not β(1 ⇒ 4) links; consequently amylose, but not cellulose, is a source of food. Cellulose is the most abundant organic compound on this planet; about 50% of organic mass is cellulose. Cotton is pure cellulose. The b glycosidic link forces each glucose to be flipped 180° with respect to the preceding but the overall chain is essentially linear. Chains then assemble first into 2D sheets which can stack into three dimensional bundles that give cotton its strength. The interactions between chains and between sheets are stabilized by hydrogen bonds.

Branched Polysaccharides.

Glycogen (Animal Starch) is an edible, highly branched

polymer (MW ~1 million, ~5000 glucose units) composed of linear segments of α (1⇒ 4) glucose (o) which branch at about every tenth glucose residue by virtue of α (1 ⇒ 6) links (•).

Complete degradation in the intestine depends on the combined action of α(1⇒ 4) amylase and an α(1⇒ 6) amylase Within the liver cell degradation is accomplished by the cooperation of glycogen phosphorylase (which cleaves the 1 ⇒ 4 bonds) plus a debranching enzyme which clips out the branch points in the "limit dextrin" produced by the enzyme phosphorylase.

- *Amylopectin* resembles glycogen except that the branches are further apart, about every 25 residues on average.
- *Plant starch,* an important constituent of our diet, is about 25% amylose and 75% amylopectin.
- Dextran is a linear polysaccharide with α(1 ⇒ 6) glycosidic links. Dextrans that have been *chemically* crosslinked (e.g. with epichlorhydrin) yield Sephadex and Biogel, the column materials used in gel filtration.
- Polysaccharides rather than monosaccharides are used as storage forms because osmotic pressure is proportional to the number of molecules and a polymer of n monosaccharides has 1/n'th the osmotic pressure of its n component sugars.

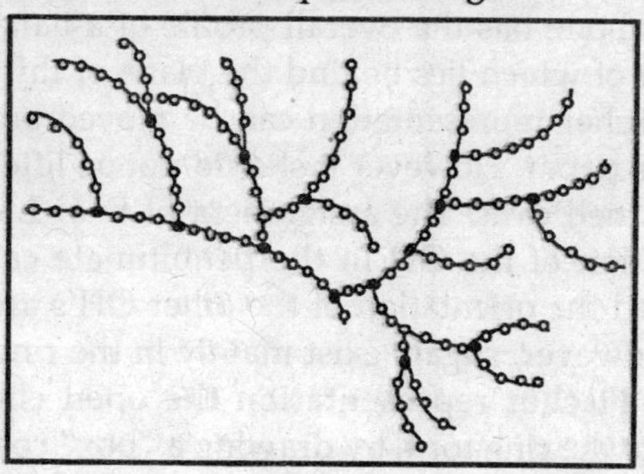

Modified Sugars

Modified sugars are important constituents of complex

carbohydrates, glycoproteins and glycolipids. They are commonly of two kinds, deoxy and amino. In deoxy sugars an -OH is replaced by -H or a $-CH_2OH$ by CH_3.

The former typically occurs at C_2 and the latter at C_6. You have already encountered 2-deoxyribose. In amino sugars the –OH is replaced by $-NH_2$, commonly at C_2. The amino group may subsequently be acetylated to give the N-acetyl derivative.

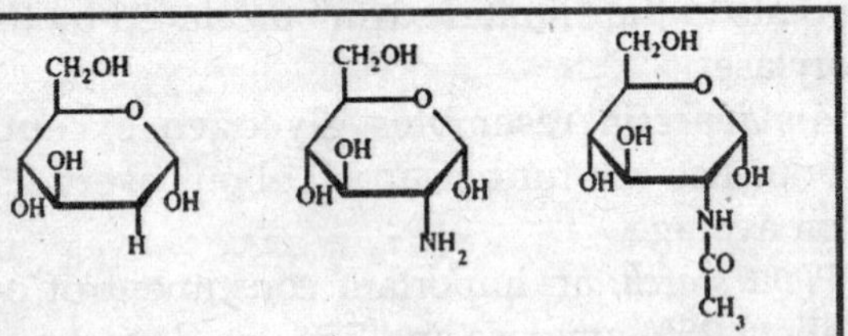

deoxy-sugar amino-sugar N-acetyl amino sugar

Representations of Sugar Structure

There are several ways to draw the structure of glucose. The simplest is the method introduced by Emil Fischer which represents the sugar as a straight chain of carbon atoms with the lowest numbered at the top and the OH's of the secondary alcohol functions shown to the right or left.

The horizontal lines are to be visualized as projecting out of the page; the vertical lines project into the page so that the carbon backbone has the overall profile of a banana, the top and bottom of which lies behind the plane of this page.

The Fischer representation can be moved around in the plane of the paper. However it should not be lifted out of the plane or turned over. The assignment of D or L depends on the orientation of the OH in the penultimate carbon (C_5 in hexoses) and the orientation of the other OH's are relative to this one. However, sugars exist mainly in the ring form.

In the Fischer representation the open chain form is converted to the ring form by drawing a "box" connecting the C atom bearing the keto function (C_1 or C_2) to C_5. If the keto function is at C_1 the $C_1 \rightarrow C_5$ link yields the pyranose ring; the $C_2 \rightarrow C_5$ link gives the furanose ring. The box is drawn on the side of the main chain as specified by the orientation of the OH at C_5 (or C_4). C1 is now chiral; if the OH created at C_1 is cis to the OH at C_5 we have the α anomeric form, and vice versa. A popular alternative is the Haworth representation which attempts to convey more three-dimensional information. The sugar is drawn as a hexagon (or pentagon), and you must imagine that the hexagon is projecting out of (and behind) the plane of the page.

This is sometimes facilitated by thickening the lower 3 edges of the hexagon. The OH's are then explicitly shown and lie either above or below the plane of the hexagon - as indicated. In the drawing the anomeric carbon and the orientation for α are indicated. However in this Haworth representation defining the sense of the anomeric carbon is not easy; the Fischer representation is better for this purpose. Three representations of glucose are:

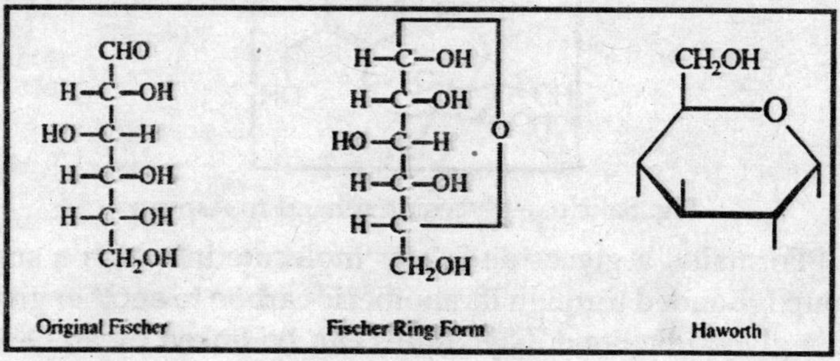

The correct configuration of the groups on the ring can be determined using three rules:

- If the ring closes on a hydroxyl which is on the right in the Fischer projection, the hydroxymethyl group (tail) points up; if it closes on a hydroxyl which was on the left in the Fischer projection, the tail points down.
- The ring hydroxyls point down if they are on the right in the Fischer projection, and up if they are on the left in the Fischer projection. (Note that these are also the positions they are in after the Fischer projection has been rotated 90° clockwise to lie on its side.)

- The hydroxyl on the anomeric carbon points down in the Δ series if it is α and up if β. In the L series, a is up and β is down.

GLYCOSIDE

In chemistry, glycosides are molecules in which a sugar is bound to a non-carbohydrate moiety, usually a small organic molecule. Glycosides play numerous important roles in living organisms. Many plants store chemicals in the form of inactive glycosides. These can be activated by enzyme hydrolysis, which causes the sugar part to be broken off, making the chemical available for use. Many such plant glycosides are used as medications. In animals (including humans), poisons are often bound to sugar molecules as part of their elimination from the body.

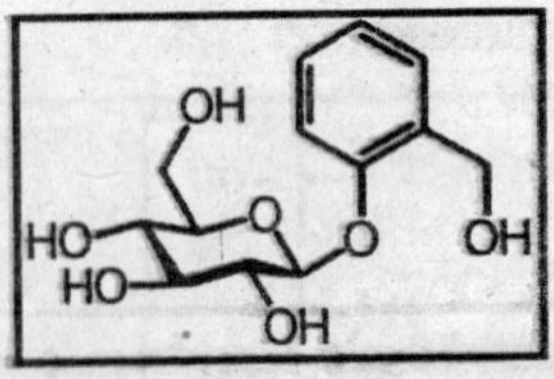

Fig. Salicin, a glycoside related to Aspirin

Formally, a glycoside is any molecule in which a sugar group is bonded through its anomeric carbon to another group via a glycosidic bond. Glycosides can be linked by an O- (an *O-glycoside*), N- (a *glycosylamine*), S-(a *thioglycoside*) or C- (a *C-glycoside*) glycosidic bond. The given definition is the one used by IUPAC.

Many authors require in addition that the sugar be bonded to a *non-sugar* for the molecule to qualify as a glycoside, thus excluding polysaccharides. The sugar group is then known as the glycone and the non-sugar group as the aglycone or genin part of the glycoside. The glycone can consist of a single sugar group (monosaccharide) or several sugar groups (oligosaccharide).

RELATED COMPOUNDS

Molecules containing an N-glycosidic bond are known as

glycosylamines and are not discussed in this article. (Many authors in biochemistry call these compounds *N-glycosides* and group them with the glycosides; this is considered a misnomer and discouraged by IUPAC.)

CHEMISTRY

Much of the chemistry of glycosides is explained in the article on glycosidic bonds. For example, the glycone and aglycone portions can be chemically separated by hydrolysis in the presence of acid. There are also numerous enzymes that can form and break glycosidic bonds.

The most important cleavage enzymes are the glycoside hydrolases, and the most important synthetic enzymes in nature are glycosyltransferases. Genetically altered enzymes termed glycosynthases have been developed that can form glycosidic bonds in excellent yield.

There are a great many ways to chemically synthesize glycosidic bonds. Fischer glycosidation refers to the synthesis of glycosides by the reaction of unprotected monosaccharides with alcohols (usually as solvent) in the presence of a strong acid catalyst. The Koenigs-Knorr reaction is the condensation of glycosyl halides and alcohols in the presence of metal salts such as silver carbonate or mercuric oxide.

CLASSIFICATION

We can classify glycosides by the glycone, by the type of glycosidic bond, and by the aglycone.

By Glycone

If the glycone group of a glycoside is glucose, then the molecule is a glucoside; if it is fructose, then the molecule is a fructoside; if it is glucuronic acid, then the molecule is a glucuronide; etc. In the body, toxic substances are often bonded to glucuronic acid to increase their water solubility; the resulting glucuronides are then excreted.

By Type of Glycosidic Bond

Depending on whether the glycosidic bond lies "below"

or "above" the plane of the cyclic sugar molecule, glycosides are classified as á-glycosides or â-glycosides. Some enzymes such as á-amylase can only hydrolyze á-linkages; others, such as emulsin, can only affect â-linkages.

By Aglycone

Glycosides are also classified according to the chemical nature of the aglycone. For purposes of biochemistry and pharmacology, this is the most useful classification.

Alcoholic Glycosides

An example of an alcoholic glycoside is salicin which is found in the genus *salix*. Salicin is converted in the body into salicylic acid, which is closely related to aspirin and has analgesic, antipyretic and antiinflammatory effects.

Anthraquinone Glycosides

These glycosides contain an aglycone group that is a derivative of anthraquinone. They have a laxative effect. They are mainly found in dicot plants except the Liliaceae family which are monocots. They are present in senna, rhubarb and *Aloe* species. Antron and anthranol are reduced forms of anthraquinone.

Coumarin Glycosides

Here the agl is coumarin or a derivative. An example is apterin which is reported to dilate the coronary arteries as well as block calcium channels. Other coumarin glycosides are obtained from dried leaves of *Psoralea corylifolia.*

Cyanogenic Glycosides

In this case, the aglycone contains a cyanide group. In many plants, these glycosides are stored in the vacuole but if the plant is attacked they are released and become activated by enzymes in the cytoplasm. These remove the sugar part of the molecule and release toxic hydrogen cyanide. Storing them in inactive forms in the cytoplasm prevents them from damaging the plant under normal conditions.

An example of these is amygdalin from almonds. They can also be found in the fruits (and wilting leaves) of the rose family (including cherries, apples, plums, almonds, peaches, apricots, raspberries, and crabapples). Cassava, an important food plant in Africa and South America, contains cyanogenic glycosides and therefore has to be washed and ground under running water prior to consumption. Sorghum (*Sorghum bicolour*) expresses cyanogenic glycosides in its roots and thus is resistant to pests such as rootworms (Diabrotica spp.) that plague its cousin maize (*Zea mays* L.). It was once thought that cyanogenic glycosides might have anti-cancer properties, but this idea was disproven, see Amygdalin. A recent study may also show that increasing CO_2 levels, caused by anthropogenic emissions, may result in much higher levels of cyanogenic glycoside production in Sorghum and Cassava plants, making them highly toxic and inconsumable. A doubling of CO_2 concentration was found to double the concentration of cyanogenic glycosides in the leaves.

Flavonoid Glycosides

Here the aglycone is a flavonoid. Examples of this large group of glycosides include:

- *Hesperidin (aglycone*: Hesperetin, glycone: Rutinose)
- *Naringin (aglycone*: Naringenin, glycone: Rutinose)
- *Rutin (aglycone*: Quercetin, glycone: Rutinose)
- *Quercitrin (aglycone*: Quercetin, glycone: Rhamnose)

Among the important effects of flavonoids are their antioxidant effect. They are also known to decrease capillary fragility.

Phenolic Glycosides (simple)

Here the aglycone is a simple phenolic structure. An example is arbutin found in the Common Bearberry *Arctostaphylos uva-ursi*. It has a urinary antiseptic effect.

Saponins

These compounds give a permanent froth when shaken with water. They also cause hemolysis of red blood cells.

Saponin glycosides are found in liquorice. Their medicinal value is due to their expectorant effect.

Steroidal Glycosides or Cardiac Glycosides

Here the aglycone part is a steroidal nucleus. These glycosides are found in the plant genera *Digitalis, Scilla,* and *Strophanthus*. They are used in the treatment of heart diseases e.g. congestive heart failure (historically as now recognised does not improve survivability; other agents are now preferred) and arrhythmia.

Steviol Glycosides

These sweet glycosides found in the stevia plant *Stevia rebaudiana* Bertoni have 40-300 times the sweetness of sucrose. The two primary glycosides, stevioside and rebaudioside A, are used as natural sweeteners in many countries. These glycosides have steviol as the aglycone part. Glucose or rhamnose-glucose combinations are bound to the ends of the aglycone to form the different compounds.

Thioglycosides

As the name implies (q.v. thio-), these compounds contain sulfur. Examples include sinigrin, found in black mustard, and sinalbin, found in white mustard.

COMPOUNDS CONTAINED IN GLYCOSIDES

- Glycosides are compounds containing a carbohydrate and a noncarbohydrate residue in the same molecule.
- The carbohydrate residue is attached by an acetal linkage at carbon atom 1 to a noncarbohydrate residue or Aglycone.
- The nonsugar component is known as the Aglycone. The sugar component is called the Glycone.
- If the carbohydrate portion is glucose, the resulting compound is a Glucoside.
- An example is the methyl glucoside formed when a solution of glucose in boiling methyl alcohol is treated with 0.5% HCl as a catalyst.

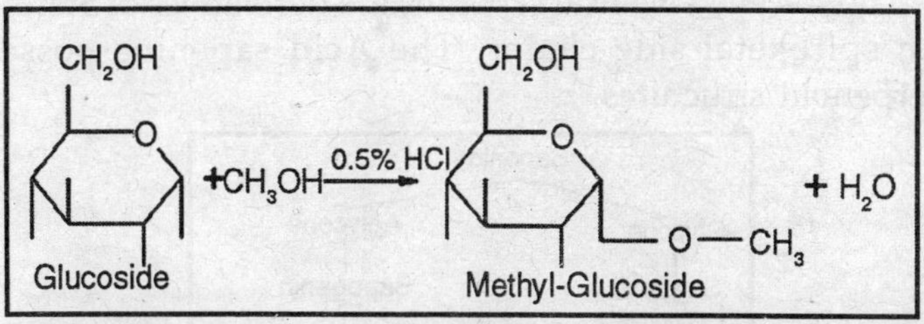

The aglycone may be methyl alcohol, glycerol, a sterol, a phenol, etc. An acetal has two ether functions at a single carbon atom.

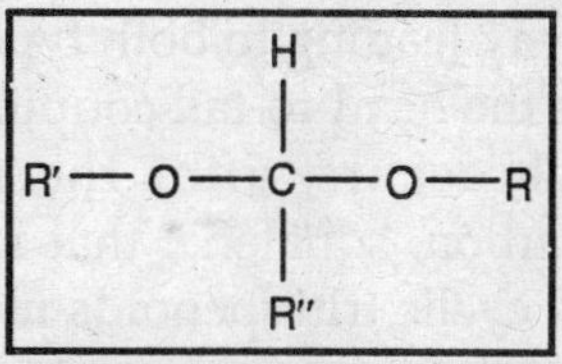

Classification of Glycosides

When the chemical nature of the aglycone group is used as the basis of systematization, the classification is as follows:

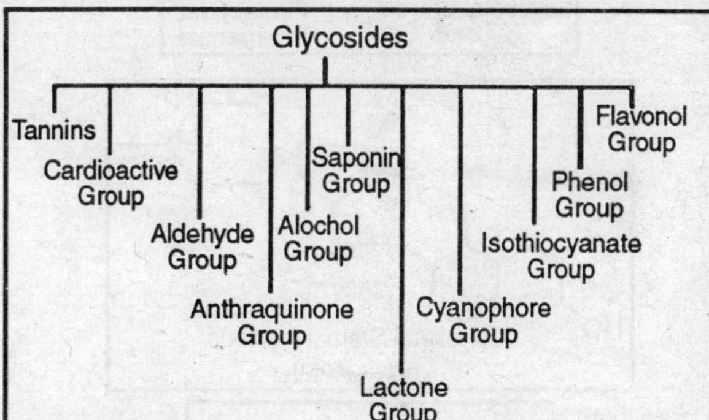

Saponin Glycosides

Saponin glycosides are divided into 2 types based on the chemical structure of their aglycones (sapogenins). Saponins on hydrolysis yield an aglycone known as "sapogenin".

Saponin = Sugar + Sapogenin

Sugar: Glycone; Sapogenin: Aglycone

The so-called Neutral saponins are derivatives of Steroids with spiroketal side chains. The Acid saponins possess triterpenoid structures.

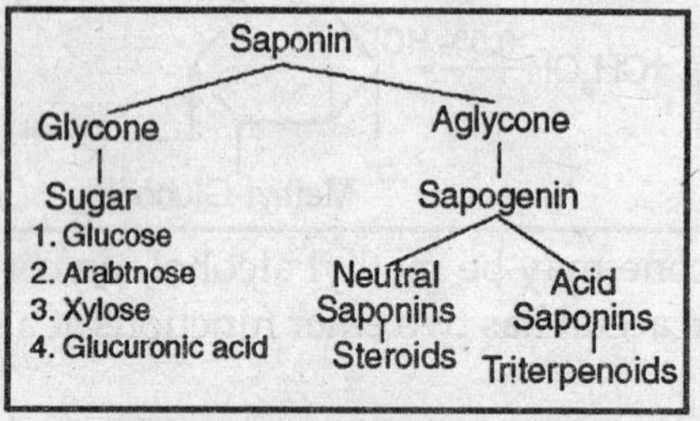

The main pathway leading to both types of sapogenins is similar and involves the head-to-tail coupling of acetate units. However, a branch occurs, after the formation of the triterpenoid hydrocarbon, squalene, that leads to steroids in one direction and to cyclic triterpenoids in the other.

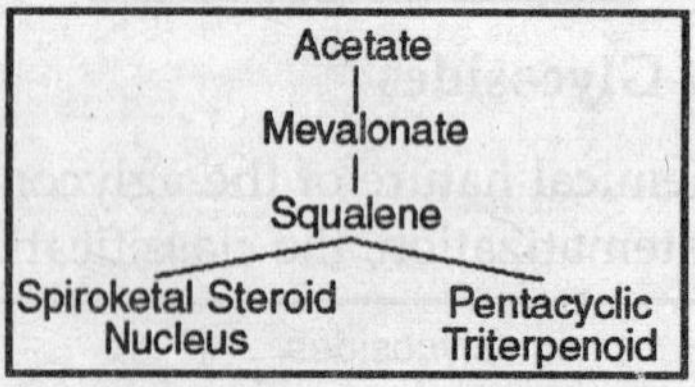

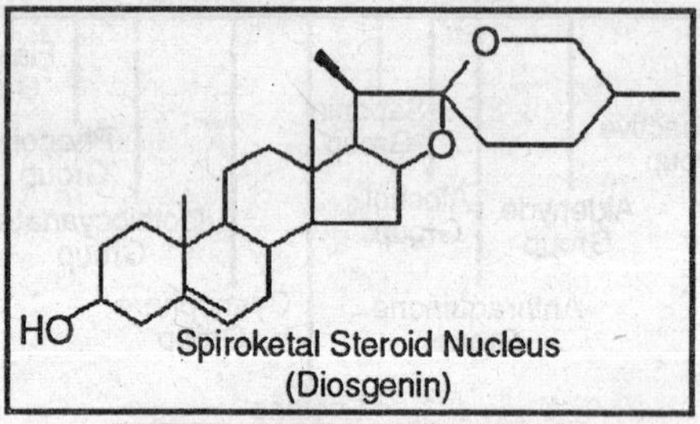

Spiroketal Steroid Nucleus
(Diosgenin)

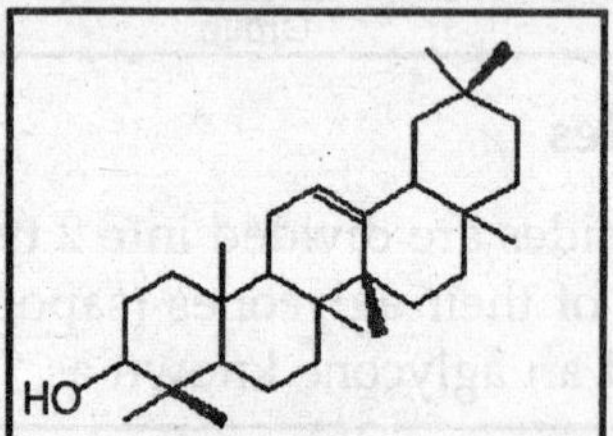

Pertacyclic Triterpenoid
(β-Amyrin)

Glycyrrhiza is the dried rhizome and roots of Glycyrrhiza glabra. *Glycyrrhiza contains:*

- A saponin glycoside called Glycyrrhizin (glycyrrhizinic acid).

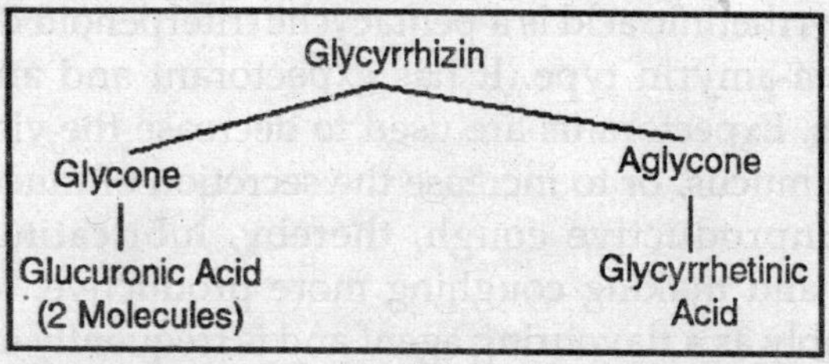

- Glycyrrhizin is the Ca^{2+} and K^+ salts of glycyrrhizinic acid.

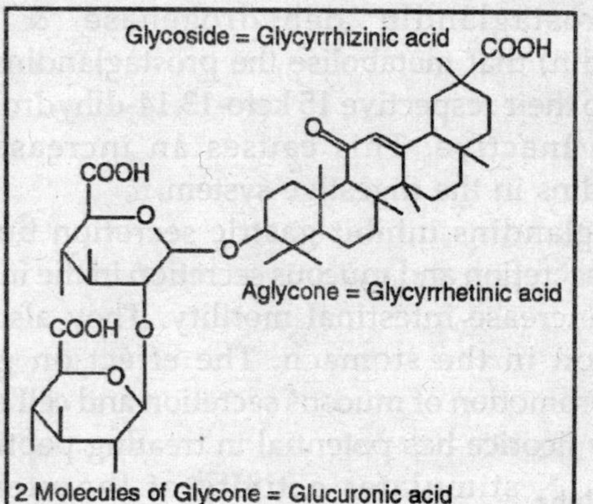

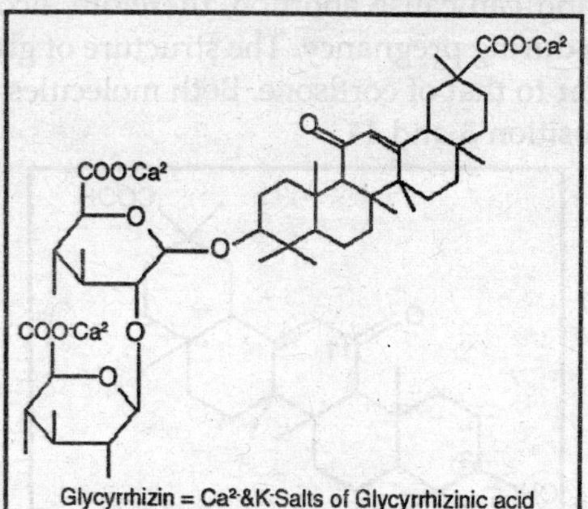

Glycyrrhizinic acid is 50 times sweeter than sugar (sucrose). Upon hydrolysis, the glycoside loses its sweet taste and is converted to the aglycone glycyrrhetinic acid plus two molecules of glucuronic acid.

Glycyrrhetinic acid is a pentacyclic triterpenoid derivative of the beta-amyrin type. It has expectorant and antitussive properties. Expectorants are used to decrease the viscosity of tenacious mucus, or to increase the secretion of mucus in dry irritant unproductive cough, thereby, lubricating the air passages and making coughing more productive. It is used considerably as a flavouring agent and is frequently employed to mask the taste of bitter drugs such as aloe, quinine etc.

Glycyrrhetinic acid inhibits the enzymes (15-hydroxyprostaglandin dehydrogenase & delta 13-prostaglandin) that metabolise the prostaglandins, PGE_2 and PGF_{2alpha} to their respective 15 keto-13,14-dihydro metabolites which are Inactive. This causes an increased level of prostaglandins in the digestive system.

Prostaglandins inhibit gastric secretion but stimulate pancreatic secretion and mucous secretion in the intestines and markedly increase intestinal motility. They also cause cell proliferation in the stomach. The effect on gastric acid secretion, promotion of mucous secretion and cell proliferation shows why licorice has potential in treating peptic ulcer.

PGF_{2alpha} stimulates activity of the uterus during pregnancy and can cause abortion, therefore, licorice should not be taken during pregnancy. The structure of glycyrrhetinic acid is similar to that of cortisone. Both molecules are flat and similar at position 3 and 11.

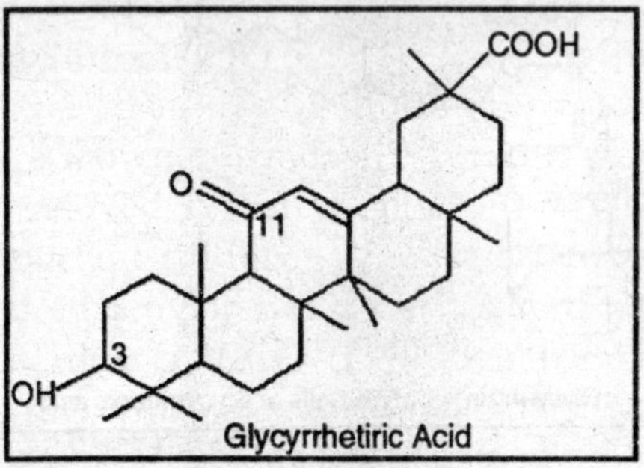

Glycyrrhetiric Acid

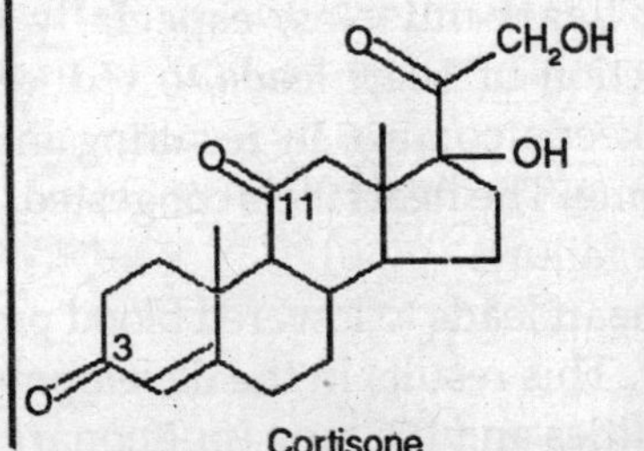

Cortisone

This might be the basis for licorice's anti-inflammatory action. 3-Beta-D-(monoglucuronyl)18-beta-glycyrrhetinic acid, a metabolite of glycyrrhetinic acid inhibits 11-beta-hydroxysteroid dehydrogenase which converts Active cortisol to Inactive cortisone in the kidneys.

The increased amounts of cortisol binds to the unprotected, unspecific mineralocorticoid receptors and induce sodium and fluid retention, hypokalaemia, hypertension and inhibition of the Renin-angiotensin-aldosterone system. Licorice should not be given to patients with a known history of high blood pressure. A prolonged usage of licorice can also cause the formation of cataracts.

Glycyrrhizin inhibits liver cell injury caused by many chemicals and is used in the treatment of chronic hepatitis and cirrhosis in Japan. It also inhibits the growth of several DNA and RNA viruses, inactivating herpes simplex virus particles irreversibly.

CARDIAC GLYCOSIDES

CONGESTIVE HEART FAILURE

Heart diseases can be primarily grouped into three major disorders: cardiac failure, ischemia and cardiac arrhythmia.

Cardiac failure can be described as the inability of the heart to pump blood effectively at a rate that meets the needs of the metabolizing tissues. This occurs when the muscles that perform contraction and force the blood out of heart are performing weakly.

Thus cardiac failures primarily arise from the reduced

contractility of heart muscles, especially the ventricles. Reduced contraction of heart leads to reduced heart output but new blood keeps coming in resulting in the increase in heart blood volume. The heart feels congested. Hence the term congestive heart failure.

Congested heart leads to lowered blood pressure and poor renal blood flow. This results in the development of edema in the lower extremities and the lung (pulmonary edema) as well as renal failure.

Cardiac Glycosides

Increasing the force of contraction of the heart (positive inotropic activity) is very important for most heart failure patients. There are several mechanisms by which this could be achieved. Cardiac steroids are perhaps the most useful and are being discussed here.

Phosphodiesterase inhibitors, such as amrinone and milrinone, have also been explored and so are direct adenylate cyclase stimulants, such as forskolin. These drugs all act by affecting the availability of intracellular Ca^{+2} for myocardial contraction or increasing the sensitivity of myocardial contractile proteins.

The cardiac glycosides are an important class of naturally occurring drugs whose actions include both beneficial and toxic effects on the heart. Plants containing cardiac steroids have been used as poisons and heart drugs at least since 1500 B.C.

Throughout history these plants or their extracts have been variously used as arrow poisons, emetics, diuretics, and heart tonics. Cardiac steroids are widely used in the modern treatment of congestive heart failure and for treatment of atrial fibrillation and flutter. Yet their toxicity remains a serious problem.

Structure

Cardiac glycosides are composed of two structural features: the sugar (glycoside) and the non-sugar (aglycone - steroid) moieties.

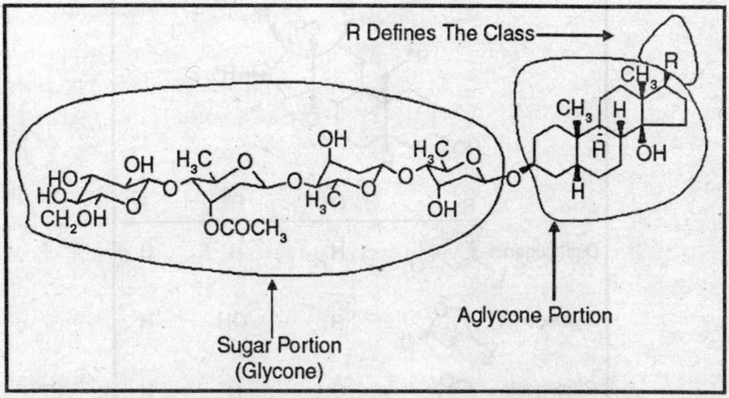

The R group at the 17-position defines the class of cardiac glycoside. Two classes have been observed in Nature - the cardenolides and the bufadienolides.

The cardenolides have an unsaturated butyrolactone ring while the bufadienolides have an a-pyrone ring.

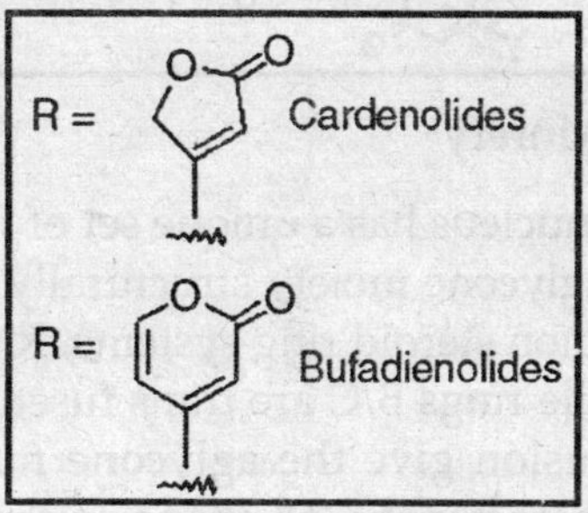

Nomenclature

The cardiac glycosides occur mainly in plants from which the names have been derived. Digitalis purpurea, Digitalis lanata, Strophanthus grtus, and Strophanthus kombe are the major sources of the cardiac glycosides.

The term 'genin' at the end refers to only the aglycone portion (without the sugar). Thus the word digitoxin refers to a agent consisting of digitoxigenin (aglycone) and sugar moieties (three).

The aglycone portion of cardiac glycosides is more important than the glycone portion.

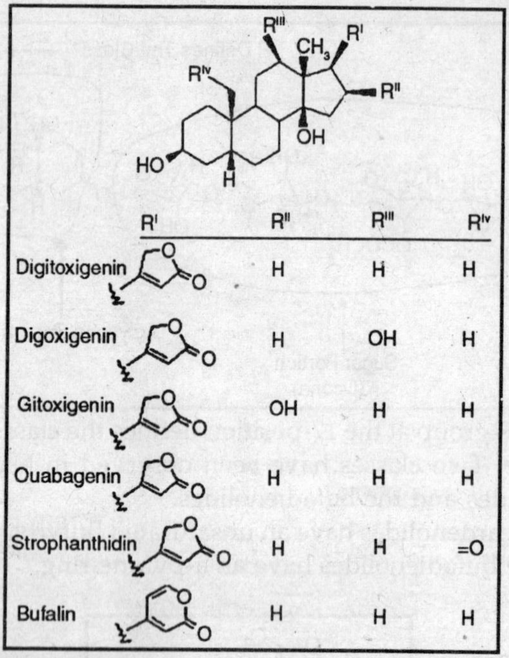

The Aglycone Moiety

The steroid nucleus has a unique set of fused ring system that makes the aglycone moiety structurally distinct from the other more common steroid ring systems. Rings A/B and C/D are cis fused while rings B/C are trans fused.

Such ring fusion give the aglycone nucleus of cardiac glycosides the characteristic 'U' shape as shown below.

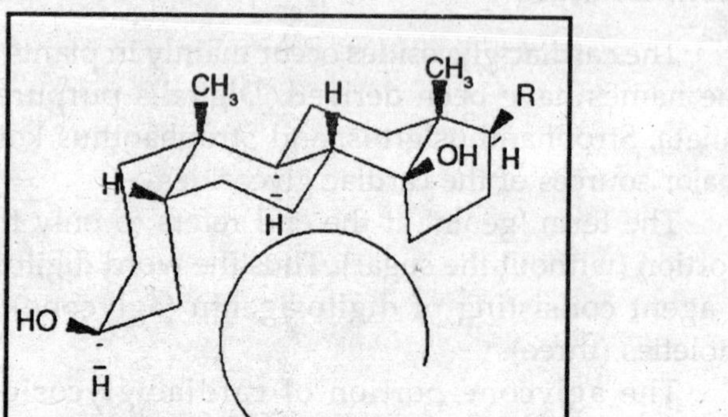

The steroid nucleus has hydroxyls at 3- and 14- positions of which the sugar attachment uses the 3-OH group. 14-OH is normally unsubstituted. Many genins have OH groups at 12- and 16- positions. These additional hydroxyl groups influence the partitioning of the cardiac glycosides into the aqueous media and greatly affect the duration of action.

The lactone moiety at C-17 position is an important structural feature. The size and degree of unsaturation varies with the source of the glycoside. Normally plant sources provide a 5-membered unsaturated lactone while animal sources give a 6-membered unsaturated lactone.

Sugar Moiety

One to 4 sugars are found to be present in most cardiac glycosides attached to the 3β-OH group. The sugars most commonly used include L-rhamnose, D-glucose, D-digitoxose, D-digitalose, D-digginose, D-sarmentose, L-vallarose, and D-fructose. These sugars predominantly exist in the cardiac glycosides in the b-conformation.

The presence of acetyl group on the sugar affects the lipophilic character and the kinetics of the entire glycoside. Because the order of sugars appears to have little to do with biological activity Nature has synthesized a repertoire of numerous cardiac glycosides with differing sugar skeleton but relatively few aglycone structures.

Structure - Activity Relationships

- The sugar moiety appears to be important only for the partitioning and kinetics of action. It possesses no biological activity. For example, elimination of the aglycone moiety eliminates the activity of alleviating symptoms associated with cardiac failure.
- The "backbone" U shape of the steroid nucleus appears to be very important. Structures with C/D *trans* fusion are inactive.
- Conversion to A/B *trans* system leads to a marked drop in activity. Thus although not mandatory A/B *cis* fusion is important.

- The 14β-OH groups is now believed to be dispensible. A skeleton without 14β-OH group but retaining the C/D *cis* ring fusion was found to retain activity.
- Lactones alone, when not attached to the steroid skeleton, are not active. Thus the activity rests in the steroid skeleton.
- The unsaturated 17-lactone plays an important role in receptor binding. Saturation of the lactone ring dramatically reduced the biological activity.
- The lactone ring is not absolutely required. For example, using α,β-unsaturated nitrile (C=C-CN group) the lactone could be replaced with little or no loss in biological activity.

Pharmacokinetics of Cardiac Glycosides

The commercially available cardiac steroids differ markedly in their degree of absorption, half-life, and the time to maximal effect.

Agent	GI absorption	Onset (m)	Peak (h)	Half-life
Ouabain	Unreliable	5-10	0.5-2	21 h
Deslanoside	Unreliable	10-30	1-2	33 h
Digoxin	55-75%	15-30	1.5-5	36 h
Digitoxin	90-100%	25-120	4-12	4-6 days

Usually this is due to the polarity differences caused by the number of sugars at C-3 and the presence of additional hydroxyls on the cardenolide. Although two cardiac glycosides may differ by only one sugar residue their partition coefficients may be significantly different resulting in different pharmacokinetics. For example, lanatoside C and digoxin differ only by a glucose residue and yet the partition coefficient measured in $CHCl_3$/16% aqueous MeOH are 16.2 and 81.5, respectively.

Glycoside	Partition Coefficient
Lanatoside C (glucose-3-acetyldigitoxose -digitoxose$_2$-digoxigenin)	16.2

Digoxin (digitoxose$_3$-digoxigenin)	81.5
Digitoxin (digitoxose$_3$-digitoxigenin)	96.5
Acetyldigoxin (3-acetyldigitoxose-digitoxose$_2$ -digoxigenin)	98.0
G-Strophanthin (rhamnose-ouabagein)	very low

In general, cardiac glycosides with more lipophilic character are absorbed faster and exhibit longer duration of action as a result of slower urinary exretion rate. Lipophilicity is markely influenced by the number of sugar residues and the number of hydroxyl groups on the aglycone part of the glycoside. Comparison of digitoxin and digoxin structures reveals that they differ only by an extra OH group in digoxin at C-12, yet their partition coefficients differ by as much as 15 % points.

Biochemical Mechanism of Action

The mechanism whereby cardiac glycosides cause a positive inotropic effect and electrophysiologic changes is still not completely clear.

Several mechanisms have been proposed, but the most widely accepted involves the ability of cardiac glycosides to inhibit the membrane bound Na^+-K^+-ATPase pump responsible for Na^+-K^+ exchange. The process of muscle contraction can be pictured as shown below.

K^+, in and out of the cell. At the resting stage, the concentration of Na^+ is high on the outside. On membrane depolarization sodium fluxes-in leading to an immediate elevation of the action potential.

Elevated intracellular Na^+ triggers the influx of free of Ca^{++} that occurs more slowly. The higher intracellular [Ca^{++}] results in the efflux of K^+. The reestablishment of the action potential occurs later by the reverse of the Na^+-K^+ exchange.

The Na^+/K^+ exchange requires energy which is provided by an enzyme Na^+-K^+-ATPase. Cardiac glycosides are proposed to inhibit this enzyme with a net result of reduced sodium exchange with potassium that leaves increased intracellular Na^+.

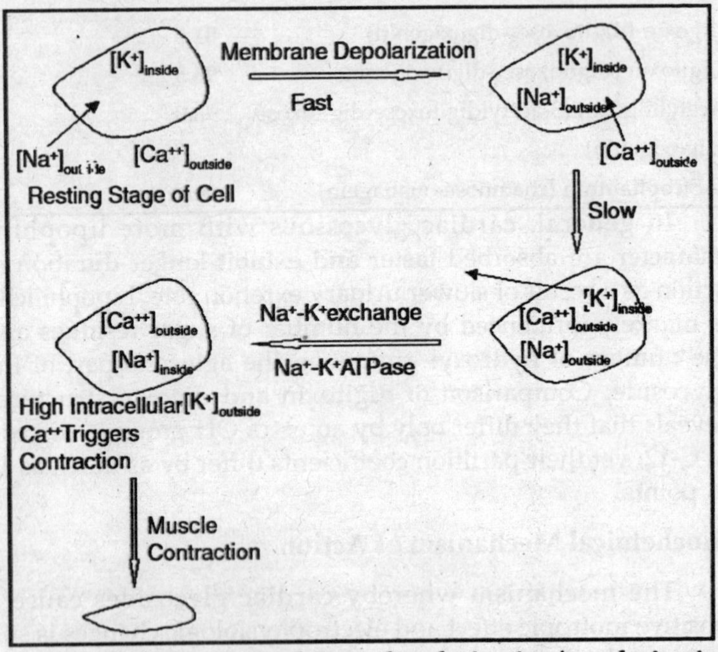

The process of membrane depolarization/repolarization is controlled by the movement of three cations, Na^+, Ca^{++}, and

This results in increased intracellular [Ca^{++}]. Elevated intracellular calcium concentration triggers a series of intracellular biochemical events that ultimately result in an increase in the force of the myocardial contraction or a positive inotropic effect.

SAPONIN

Saponins are a class of chemical compounds, one of many secondary metabolites found in natural sources, with saponins found in particular abundance in various plant species.

Specifically, they are amphipathic glycosides grouped phenomenologically by the soap-like foaming they produce when shaken in aqueous solutions, and structurally by their being composed of one or more hydrophilic glycoside moieties combined with a lipophilic triterpene derivative. A ready and therapeutically relevant example is the cardio-active agent digoxin, from common foxglove.

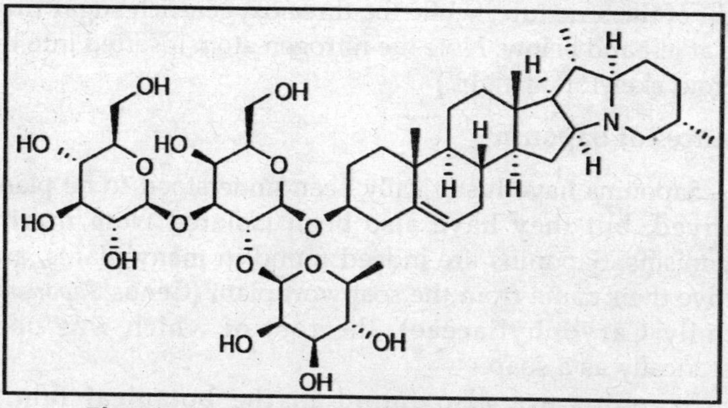

Fig. Chemical Structure of the Saponin Solanine

Structural Variety and Biosynthesis

The aglycone (glycoside-free portion) of the saponins are termed sapogenins. The number of saccharide chains attached to the sapogenin/aglycone core can vary – giving rise to another dimension of nomenclature (monodesmosidic, bidesmosidic, etc.) – as can the length of each chain.

A somewhat dated compilation has the range of saccharide chain lengths being 1–11, with the numbers 2-5 being the most frequent, and with both linear and branched chain saccharides being represented. Dietary monosaccharides such as D-glucose and D-galactose are among the most common components of the attached chains.

The lipophilic aglycone can be any one of a wide variety of polycyclic organic structures originating from the serial addition of ten-carbon (C10) terpene units to compose a C30 triterpene skeleton, often with subsequent alteration to produce a C27 steroidal skeleton.

The subset of saponins that are steroidal have been termed saraponins; Aglycone derivatives can also incorporate nitrogen, so that some saponins also present chemical and pharmacologic characteristics of alkaloid natural products.

The figure at right above presents the structure of the alkaloid phytotoxin solanine, a monodesmosidic, branched-saccharide steroidal saponin. (The lipophilic steroidal structure is the series of connected six- and five-membered rings at the right of the structure, while the three oxygen-rich sugar rings are at left and below. Note the nitrogen atom inserted into the steroid skeleton at right.)

Sources of Saponin

Saponins have historically been understood to be plant-derived, but they have also been isolated from marine organisms. Saponins are indeed found in many plants, and derive their name from the soapwort plant (Genus *Saponaria*, Family Caryophyllaceae), the root of which was used historically as a soap.

Saponins are also found in the botanical family Sapindaceae, with its defining genus *Sapindus* (soapberry or soapnut), and in the closely related families Aceraceae (maples) and Hippocastanaceae (horse chestnuts; ref. needed). It is also found heavily in gynostemma pentaphyllum (Genus *Gynostemma*, Family Cucurbitaceae) in a form called gypenosides, and ginseng (Genus *Panax*, Family Araliaceae) in a form called ginsenosides.

Within these families, this class of chemical compounds are found in various parts of the plant: leaves, stems, roots, bulbs, blossom, and fruit.

Commercial formulations of plant-derived saponins – e.g., from the soap bark (or soapbark) tree, *Quillaja saponaria*, and from other sources—are available via controlled manufacturing processes, which make them of use as chemical and biomedical reagents.

Role in Plant Ecology and Impact on Animal Foraging

In plants, saponins may serve as anti-feedants, and to protect the plant against microbes and fungi. Some plant saponins (e.g. from oat and spinach) may enhance nutrient absorption and aid in animal digestion.

However, saponins are often bitter to taste, and so can reduce plant palatability (e.g., in livestock feeds), or even imbue them with life-threatening animal toxicity.

Data make clear that some saponins are toxic to cold-blooded organisms and insects at particular concentrations.

There is a need for further research to define the roles of these natural products in their host organisms—which have been described as "poorly understood" to date.

Established Research Bioactivities and Therapeutic Claims

Bioactivities

One research use of the saponin class of natural products involves their complexation with cholesterol to form pores in cell membrane bilayers, e.g., in red cell (erythrocyte) membranes, where complexation leads to red cell lysis (hemolysis) on intravenous injection.

In addition, the amphipathic nature of the class gives them activity as surfactants that can be used to enhance penetration of macromolecules such as proteins through cell membranes. Saponins have also been used as adjuvants in vaccines.

Medical Uses

There is tremendous, commercially driven promotion of saponins as dietary supplements and nutriceuticals. There is evidence of the presence of saponins in traditional medicine preparations, where oral administrations might be expected to lead to hydrolysis of glycoside from terpenoid (and obviation of any toxicity associated with the intact molecule). But as is often the case with wide-ranging commercial therapeutic claims for natural products:

- The claims for organismal/human benefit are often based on very preliminary biochemical or cell biological studies; and
- Mention is generally omitted of the possibilities of individual chemical sensitivity, or to the general toxicity of specific agents,) and high toxicity of selected cases.

While such statements require constant review, it appears that there are very limited US, EU, etc. agency-approved roles for saponins in human therapy. In their use as adjuvants in the production of vaccines, toxicity associated with sterol complexation remains a major issue for attention. Even in the case of digoxin, therapeutic benefit from the cardiotoxin is a result of careful administration of an appropriate dose.

Very great care needs to be exercised in evaluating or acting on specific claims of therapeutic benefit from ingesting saponin-type and other natural products.

Triterpenoid Saponins

Triterpenoid saponins are triterpenes which belong to the group of saponin compounds.

Triterpenes belong to a large group of compounds arranged in a four or five ring configuration of 30 carbons with several oxygens attached.

Triterpenes are assembled from a C5 isoprene unit through the cytosolic mevalonate pathway to make a C30 compound and are steroidal in nature.

Cholesterol is one example of a triterpene. Phytosterols and phytoecdysteroids are also triterpenes. The triterpenes are subdivided into some 20 groups, depending on their particular structures.

Though all terpenoid compounds have bioactivity in mammals, it is the triterpenes that are most important to the adaptogenic effect found in plants such as Panax ginseng or Eleutherococcus senticosus.

Most triterpenoid compounds in adaptogenic plants are found as saponin glycosides which refers to the attachment of various sugar molecules to the triterpene unit. These sugars can be easily cleaved off in the gut by bacteria, allowing the aglycone (triterpene) to be absorbed.

This allows them insert into cell membranes and modify the composition, influence membrane fluidity, and potentially affect signaling by many ligands and cofactors. Saponin glycosides reduce surface tension of water with foaming and will break down lipids.

Usually triterpene saponins are designated as such by the suffix ending –side, such as ginsenoside, or astragaloside, named for the plant genera they were first discovered in.

Some, such as the ginsenosides and eleutherosides are

designated Rx where the suffix x = a, a1, b2, is indicative of the relative position of the saponin spots from top to bottom of a thin layer chromatogram.

Common Medicinally Active Triterpenoid Saponins From Adaptogenic Plants

Chemical compound: Common Name (Latin Name)

- *20-hydroxyecdysone*: Maral root (Rhaponticum carthamoides)
- *Araloside*: Spikenard (Aralia mandshurica)
- *Astragaloside*: huang qi (Astragalus membranaceus}
- *Bacoside*: Brahmi (Bacopa monniera)
- *Cucurbitacin*: Bryonia (Bryonia alba)
- *Eleutheroside*: Siberian ginseng (Eleutherococcus senticosus)
- *Ginsenoside*: Ginseng (Panax ginseng)
- *Gypenoside*: Jiaogulan (Gynostemma pentaphyllum)
- *Tanshenoside*: Bellflower (Codonopsis pilosula)
- *Tinosporoside*: Guruchi (Tinospora cordifolia)
- *Withanolide*: Ashwagandha (Withania somnifera)

Steviol Glycoside

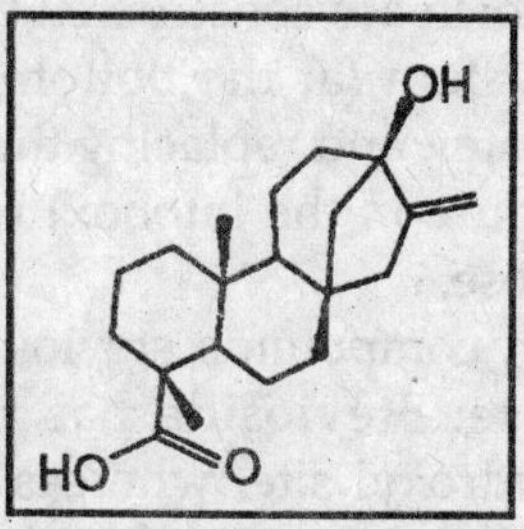

Fig. Steviol

The steviol glycosides are responsible for the sweet taste of the leaves of the stevia plant (*Stevia rebaudiana* Bertoni). These compounds range in sweetness from 40 to 300 times sweeter than sucrose.

They are heat-stable, pH-stable, and do not ferment. They also do not induce a glycemic response when ingested, making them attractive as natural sweeteners to diabetics and others on carbohydrate-controlled diets.

Structure

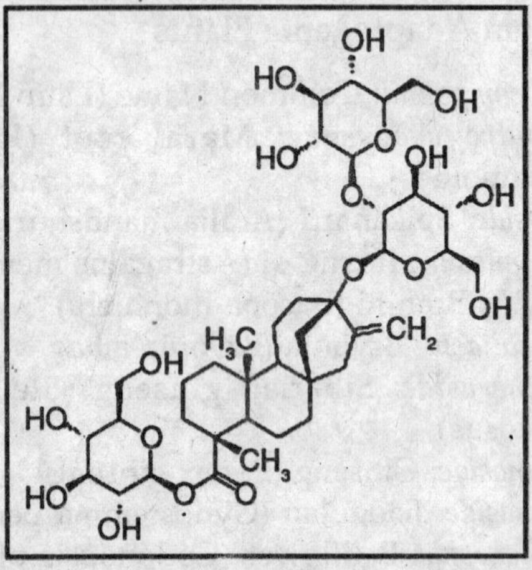

Fig. Stevioside, Formed by Substituting Two Linked Glucose Molecules For the Top Hydrogen Atom and One Glucose Molecule for the Bottom Hydrogen.

The diterpene known as Steviol is the aglycone of stevia's sweet glycosides, which are constructed by replacing steviol's carboxyl hydrogen atom (at the bottom of the figure) with glucose to form an ester, and replacing the hydroxyl hydrogen (at the top of the figure in the infobox) with combinations of glucose and rhamnose.

The two primary compounds, stevioside and rebaudioside A, use only glucose: Stevioside has two linked glucose molecules at the hydroxyl site, whereas rebaudioside A has three, with the middle glucose of the triplet connected to the central steviol structure.

In terms of weight fraction, the four major steviol glycosides found in the stevia plant tissue are:

- 5–10% stevioside
- 2–4% rebaudioside A — most sweet and least bitter
- 1–2% rebaudioside C
- ½–1% dulcoside A.

Rebaudioside B, D, and E may also be present in minute quantities; however, it is suspected that rebaudioside B is a byproduct of the isolation technique. The two majority compounds stevioside and rebaudioside, primarily responsible for the sweet taste of stevia leaves, were first isolated by two French chemists in 1931.

STEVIOL GLYCOSIDES

Chemical and Technical Assessment

Water extracts of the crushed leaves of the stevia plant, the perennial shrub *Stevia rebaudiana* (Bertoni) Bertoni of the family *Asteraceae (Compositae),* have been used for many years as a sweetener in some countries in South America and Asia.

In the United States, however, although the leaves of the Stevia plant or extracts have been marketed as dietary ingredients in dietary supplements, these substances currently are not permitted as conventional food ingredients or additives with a sweetener function.

The composition of the extracts depends on the composition of the leaves, influenced by soil and climate, and on the extraction and purification processes used. The impurities occurring in extracts of the Stevia leaves are typical plant materials, such as pigments and saccharides.

Submission from countries in different parts of the world suggest that the main components of the commercially available extracts of stevia contain as the main components stevioside and rebaudioside A in various amounts ranging from about 10-70% stevioside and 20-70% rebaudioside A.

It appears that most commercial products have a total steviol glycoside content of more than 90% with the the two main steviol glycosides making up about 80% of the material.

Nomenclature

As described above, extracts of *Stevia* have variable compositions. Structures of stevioside and related compounds are given in Figure. Steviol is the aglycone of all of the principle and secondary sweetener components (the glycoside terpenes).

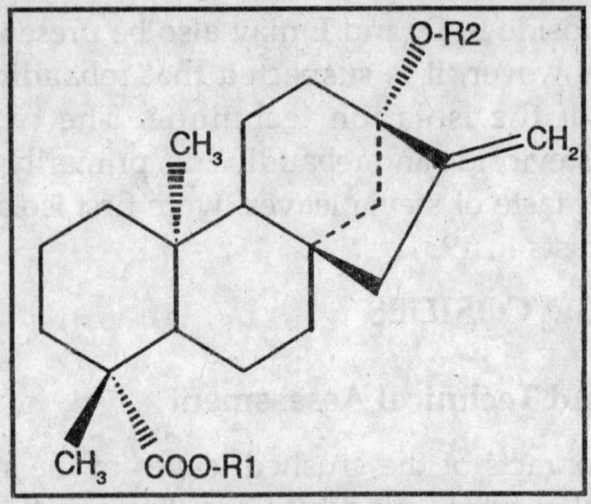

Fig. Structures of Stevioside.

Table. Related Compounds of Stevioside.

Compound name	R1	R2
Steviol	H	H
Steviolbioside	H	β-Glc-β-Glc(2→1)
Stevioside	β-Glc	β-Glc-β-Glc(2→1)
Rebaudioside A	β-Glc	β-Glc-β-Glc(2→1)
		|
Rebaudioside B	H	β-Glc(3→1)
		β-Glc-β-Glc(2→1)
		|
Rebaudioside C	β-Glc	β-Glc(3→1)
(dulcoside B)		β-Glc-α-Rha
	(2→1)	
		|
		β-Glc(3→1)
Rebaudioside D	β-Glc-β-Glc(2→1)	β-Glc-β-Glc(2→1)
		|
		β-Glc(3→1)
Rebaudioside E	βGlc-β-Glc(2→1)	β-Glc-β-Glc(2→1)
Rebaudioside F	β-Glc	β-Glc-β-Xyl(2→1)
		|
		β-Glc(3→1)
Dulcoside A	β-Glc	β-Glc-α-Rha
		(2→1)

The commercially available products under various names (Stevioside, Stevia extracts, Purified Stevia Extract etc.) vary in composition from one manufacture of sweetener product to another with respect to the relative ratios of the steviol glycosides and other constituents. Extracts of stevia leaves manufactured with high levels of Rebaudioside A compared to levels of stevioside have been produced in Japan during the last twenty years.

Four such products, namely Stevia extract, Stevia essence, Stevia powder, and Stevia sweetener, were included in "List of Food Additives Excluding Chemical Synthetics," published in 1989 by the Ministry of Health and Welfare, Japan. There is no single common or trivial name established through common usage that is available to identify the specific commercial materials being evaluated at the current JECFA.

In 1988 the 33rd JECFA developed guidelines for designating titles for specification monographs. Based on these guidelines the Committee named the specification monograph "Steviol glycosides" to indicate that the commercial products are mixtures of the steviol glycosides.

Description: Steviol glycosides preparations are white or slightly yellowish white crystalline odourless or having a slight characteristic odour, water soluble powders, which are 200 to 300 times sweeter than sucrose.

Manufacturing

Raw Materials: The raw materials used in the manufacturing process of Steviol glycosides preparations are the crushed leaves of the perennial shrub *Stevia rebaudiana* (Bertoni) Bertoni of the family *Asteraceae (Compositae)*. Various alcohols and ion exchange resins are used in the manufacturing.

Methods of Manufacture

Manufacturer from Paraguay

Material from South America may contain greater than 95% total carbohydrate, with stevioside content around 70%

and rebaudioside A approximately 20%; minor steviol glycosides are likely not more than 5% total. The basic stages of the extraction process of sweetener principles of *Stevia rebaudiana* leaves are:

- Preparation of the aqueous extract
- Preparation of the organic extract
- Obtaining stevicl glycosides
- Preparation of the aqueous extract

The ground leaves of *Stevia rebaudiana* are mixed with hot water for 20-30 min. Subsequently, the aqueous extract is removed by draining, using pressure in order to achieve the maximum amount of extract.

Several types of infusion/draining processes may be used. The extract is allowed to cool to room temperature. In order to remove impurity particles, the extract may be allowed to rest while particulate matter settles out or it may be centrifuged. The extract contains the sweetener principles, the plant pigments and other water-soluble components.

Preparation of the organic extract: The sweetener principles and the pigments are extracted with a mixture of butanol or isobutanol and a less polar solvent, such as benzene, chloroform or hexane.

Obtaining steviol glycosides: The organic phase is separated and concentrated until a solid mass is formed. The mass is dissolved in hot methanol. Steviol glycoside crystals form on cooling.

The crystals are separated and washed with cold methanol, and finally recrystallized from methanol/water. The resulting material has a high degree of purity (97-98% steviol glycosides) and contains about 4% water.

Manufacturer from Japan

The manufacturing process used in Japan is reported to yield a product containing 80-88% of the four sweetening agents (stevioside, rebaudioside A, rebaudioside C and dulcoside A). Other steviol glycosides and saccharides are present at 5-10%.

The manufacturing process is presented as a flow sheet:

Process	Operating conditions
Dry *Stevia* leaves	Harvest; Leaves and stem dissociation
Extraction	Extract 2-3 times with water at 50-60°
Flocculation	Add calcium hydroxide and aluminium sulphate
Filtration	Filter press
Adsorption on resin	The steviol glycosides are adsorbed and then released with alcohol
Distillation	Alcohol vaporisation
Ion exchange	Cation and anion resin mix bed
Concentration	Vaporize water under reduced pressure
Micro filter	1 μm
Sterilization	120°, 1 min
Spray drying	Drying

Chemical Characterization

Composition of the Food Additive

The chemical composition of steviol glycosides preparations varies considerably depending on the manufacturing process and also on the leaves used as raw material, as their composition varies with the temperature during the year of cultivation, the length of daylight during the growth period, and the time of harvest.

Based on information, products on the market contain not less than 80% of the two principal and two secondary steviol glycosides, and the sum of the percentages of stevioside and rebaudioside A is not less than 70%. Other steviol glycosides are also present in minor amounts (5-10%). However, the composition (qualitative and quantitative) of the remaining fractions of the extracts has not been sufficiently characterized.

Possible impurities: The impurities occurring in steviol glycosides are due primarily to compounds extracted from the Stevia leaves. Analytical results on preparations support a setting cf maximum limits of 1 mg/kg for both arsenic and lead.

Analytical methods: Different methods, mainly liquid chromatographic methods, are currently available for the identification and determination of the principal steviol glycosides.

Functional Uses and Reactions/Fate in Foods

Steviol glycoside extracts are mainly used as a sweetener in the manufacture of fruit and milk drinks, desserts, yoghurt, cold confectionary, delicacies and pickles. Stevioside and Rebaudioside A are reasonably stable under the elevated temperatures used in food processing and do not undergo browning or caramelization when heated.

Use Levels in Foods

The steviol glycosides are used to sweeten a number of foods in Asia and South America.

Table. Food Use Levels Reported.

Food Type	Reported Maximum Use Level (mg/kg)
Beverages	500
Desserts	500
Yogurt	500
Cold Confectionery	500
Sauces	1000
Pickles	1000
Delicacies	1000
Sweet Corn	200
Bread	160
Biscuits	300

It is also known that stevia leaves are used to prepare a sweetened tea in a number of countries throughout the world.

Toxicity

A 1985 study reporting that steviol may be a mutagen has been criticized on procedural grounds that the data were mishandled in such a way that even distilled water would appear mutagenic. More recent studies appear to establish the safety of steviol and its glycosides.

In 2006, the World Health Organization (WHO) performed a thorough evaluation of recent experimental studies of stevia extracts conducted on animals and humans, and concluded that "stevioside and rebaudioside A are not genotoxic in vitro or in vivo and that the genotoxicity of steviol and some of its oxidative derivatives in vitro is not expressed in vivo."

The report also found no evidence of carcinogenic activity. The report also suggested the possibility of health benefits, in that "stevioside has shown some evidence of pharmacological effects in patients with hypertension or with type-2 diabetes", but concluded that further study was required to determine proper dosage.

STEVIA

Stevia is a genus of about 240 species of herbs and shrubs in the sunflower family (Asteraceae), native to subtropical and tropical South America and Central America.

The species Stevia rebaudiana, commonly known as sweetleaf, sweet leaf, sugarleaf, or simply stevia, is widely grown for its sweet leaves. As a sweetener and sugar substitute, stevia's taste has a slower onset and longer duration than that of sugar, although some of its extracts may have a bitter or licorice-like aftertaste at high concentrations.

Fig. Stevia Rebaudiana Flowers.

With its extracts having up to 300 times the sweetness of sugar, stevia has garnered attention with the rise in demand for low-carbohydrate, low-sugar food alternatives. Medical research has also shown possible benefits of stevia in treating obesity and high blood pressure. Because stevia has a negligible effect on blood glucose, it is attractive as a natural sweetener to people on carbohydrate-controlled diets.

However, health and political controversies have limited stevia's availability in many countries; for example, the United States banned it in the early 1990s unless labeled as a supplement, although in 2008 it became commercially available as a sweetener. Stevia is widely used as a sweetener in Japan, and it is available in Canada as a dietary supplement.

Use of Stevia

The genus *Stevia* consists of 240 species of plants native to South America, Central America, and Mexico, with several species found as far north as Arizona, New Mexico, and Texas. They were first researched by botanist and physician Pedro Jaime Esteve; the word *stevia* is a Latinized derivation of his surname.

Human use of the sweet species *S. rebaudiana* originated in South America. The leaves of the stevia plant have 30–45 times the sweetness of sucrose (ordinary table sugar). The leaves can be eaten fresh, or put in teas and foods.

In 1899, the Swiss botanist Moisés Santiago Bertoni, during his research in eastern Paraguay first described the plant and the sweet taste in detail. But only limited research was conducted on the topic, until in 1931, two French chemists isolated the glycosides that give stevia its sweet taste. These compounds were named stevioside and rebaudioside, and are 250–300 times sweeter than sucrose, heat stable, pH stable, and non-fermentable.

The exact structure of the aglycone and the glycoside were published in 1955. In the early 1970s, Japan began cultivating stevia as an alternative to artificial sweeteners such as cyclamate and saccharin, which were suspected carcinogens. The plant's leaves, the aqueous extract of the leaves, and purified steviosides are used as sweeteners.

Since the Japanese firm Morita Kagaku Kogyo Co., Ltd. produced the first commercial stevia sweetener in Japan in 1971, the Japanese have been using stevia in food products, soft drinks (including Coca Cola), and for table use. Japan currently consumes more stevia than any other country, with stevia accounting for 40% of the sweetener market.

Today, stevia is cultivated and used in food elsewhere in east Asia, including in China (since 1984), Korea, Taiwan, Thailand, and Malaysia. It can also be found in Saint Kitts and Nevis, in parts of South America (Brazil, Colombia, Peru, Paraguay, and Uruguay) and in Israel. China is the world's largest exporter of stevioside.

Stevia species are found in the wild in semi-arid habitats ranging from grassland to mountain terrain. Stevia does produce seeds, but only a small percentage of them germinate. Planting cloned stevia is a more effective method of reproduction.

Medicinal Use

For centuries, the Guaraní tribes of Paraguay, Bolivia and Brazil used stevia, which they called ka'a he'ê ("sweet herb"), as a sweetener in yerba mate and medicinal teas for treating heartburn and other ailments. More recent medical research has shown promise in treating obesity and hypertension. Stevia has a negligible effect on blood glucose, even enhancing glucose tolerance; therefore, it is attractive as a natural sweetener to diabetics and others on carbohydrate-controlled diets.

Fig. S. Rebaudiana Foliage.

Possible treatment of osteoporosis has been suggested by

the patent application claim that eggshell breakage can be reduced by 75% by adding a small percentage of stevia leaf powder to chicken feed. It has also been suggested that pigs fed stevia extract had twice as much calcium content in their meat, but these claims have been unverified.

Availability

Stevia has been grown on an experimental basis in Ontario, Canada since 1987 for the purpose of determining the feasibility of growing the crop commercially. In the United States, Rebiana is generally recognized as safe as of December 2008, and stevia is also recognized as a dietary supplement.

Stevia has also been approved as a dietary supplement in Australia, New Zealand and Canada. Since June 2008 it is approved as a sweetener for food and beverages in Australia and New Zealand.

In Japan and South American countries, stevia may also be used as a food additive. Stevia is currently banned for use in food in the European Union except in France which approved it for a 2-year test starting September 2009. It is also banned in Singapore and Hong Kong.

In 2007, The Coca-Cola Company announced plans to obtain approval for their stevia-derived sweetener rebiana for use as a food additive within the United States by 2009, as well as plans to market rebiana-sweetened products in 12 countries that allow stevia's use as a food additive. In May 2008, Coke and Cargill announced the availability of Truvia, a consumer brand stevia sweetener containing erythritol and Rebiana, which the FDA permitted as a food additive in December 2008. Coca-Cola announced intentions to release stevia-sweetened beverages in late December 2008.

Shortly afterward, PepsiCo and Pure Circle announced PureVia, their brand of stevia-based sweetener, but withheld release of beverages sweetened with reb-A until receipt of FDA confirmation. Since the FDA permitted Truvia and PureVia, both Coca Cola and PepsiCo have announced products that will contain their new sweetener.

Critics note that the FDA has not actually permitted the

stevia plant itself to be used as a food additive, but only the Reb-A extract.

Production

To produce rebaudioside A commercially, stevia plants are dried and subjected to a water extraction process. This crude extract contains about 50% rebaudioside A and is refined using crystallization and separation technologies to separate the various glycoside molecules in the extract. This allows the manufacturer to isolate pure rebaudioside A.

The National Research Council of Canada has patented a process for extracting sweet compounds from Stevia by column extraction at temperatures from 0-25°C, followed by purification by nanofiltration. A microfiltration pretreatment step is used to clarify the extract. Purification is by ultrafiltration followed by nanofiltration.

Steviol is the basic building block of stevia's sweet glycosides: Stevioside and rebaudioside A are constructed by replacing the bottom hydrogen atom with glucose and the top hydrogen atom with two or three linked glucose groups, respectively.

Health Controversy

A 1985 study reported that steviol, a breakdown product from stevioside and rebaudioside (two of the sweet steviol glycosides in the stevia leaf), is a mutagen in the presence of a liver extract of pre-treated rats — but this finding was criticized on procedural grounds that the data were mishandled in such a way that even distilled water would appear mutagenic.

Over the following years bioassay, cell culture, and animal studies have shown mixed results in terms of toxicology and adverse effects of stevia constituents, but in general, they have not been found to be harmful. While reports emerged that found steviol and stevioside to be weak mutagens, the bulk of studies show an absence of harmful effects.

In a 2008 review, 14 of 16 studies cited showed no genotoxic activity for stevioside, 11 of 15 studies showed genotoxic activity for steviol, and no studies showed genotoxicity for Rebaudioside A.

Nevertheless, even if a chemical can cause DNA damage in the controlled conditions of a bioassay (e.g., in bacteria, in mammalian cell cultures) it is a fundamentally different question whether it causes cancer in intact organisms (e.g., rodents, humans) or is teratogenic (i.e., causes birth defects). No evidence for stevia constituents causing cancer or birth defects has been found.

Other studies have shown stevia to improve insulin sensitivity in rats and possibly even to promote additional insulin production, helping to reverse diabetes and metabolic syndrome. Preliminary human studies suggest that stevia can help reduce hypertension although another study has shown it to have no effect on hypertension. Indeed, millions of Japanese have been using stevia for over thirty years with no reported or known harmful effects. Similarly, stevia leaves have been used for centuries in South America spanning multiple generations in ethnomedical tradition as a treatment for type II diabetes. In 2006, the World Health Organization (WHO) performed a thorough evaluation of recent experimental studies of stevioside and steviols conducted on animals and humans, and concluded that "stevioside and rebaudioside A are not genotoxic in vitro or in vivo and that the genotoxicity of steviol and some of its oxidative derivatives in vitro is not expressed in vivo." The report also found no evidence of carcinogenic activity. Furthermore, the report noted that "stevioside has shown some evidence of pharmacological effects in patients with hypertension or with type-2 diabetes" but concluded that further study was required to determine proper dosage.

Whole foods proponents draw a distinction between consuming (and safety testing) only parts, such as stevia extracts and isolated compounds like stevioside, versus the whole herb. However, professionals in pharmacognosy, as well as physicians and science journalists disagree that whole foods are beneficial over extracted components, and may even be harmful.

Chapter 6

Biosynthesis of Amino Acids

Amino acids and peptides are present in humans, animals, tissues, blood, microorganisms and plants.

MEDICAL AND BIOLOGICAL IMPORTANCE

- Amino acids serve as building blocks of proteins. Some amino acids are found in free form in human blood.
- They also serve as precursors of hormones, purines, pyrimidines, porphyrins, vitamins and biologically important amines like histamine.
- Peptides have many important biological functions. Some of them are hormones. They are used as antibiotics and antitumor agents.
- Some peptides are required for detoxification reactions. Some peptides serve as neurotransmitters.
- Amino acid proline protects living organisms against free radical induced damage.
- Some peptides are involved in regulation of cell cycle and apoptosis.
- Peptides of vertebrates and invertebrates act as antimicrobial agents. They are part of innate immunity. Bacterial infections at epithelial surface induce production of antimicrobial peptides, which cause lysis of microbes.
- Peptides are enzyme inhibitors. Natural and synthetic peptide inhibitors of angiotensin converting enzyme

(ACE) act as a anti hypertensives. Peptide inhibitors of ACE present in physiological foods, lowers blood pressure after they are absorbed from intestine. Lisinopril, Enalapril etc. are synthetic peptide inhibitors of ACE that are used as drugs in the treatment of hypertension.

- Some synthetic peptides are used as enzyme substrates.

CHEMICAL NATURE OF AMINO ACIDS

Amino acids are carboxylic acids containing an amino group. In most of the amino acids, an amino group is attached to α-carbon atom next to the carboxyl group hence they are α-aminoacids. The general formula is shown in Figure.

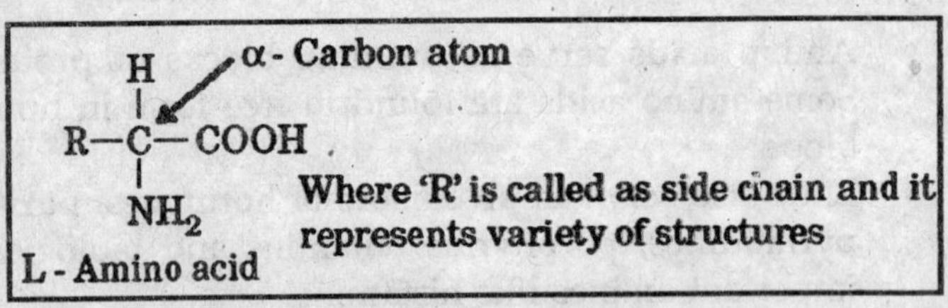

Fig. Structure of an α-Amino Acid

COMMON AMINO ACIDS

Though more than 200 amino acids are identified in nature, only 20 amino acids serve as building blocks of body proteins. They are known as common amino acids. In addition to the common amino acids, derived amino acids are also found in proteins.

CLASSIFICATION OF AMINO ACIDS

Amino acids have been classified in various ways.

I. Based on side chain and ring structure present, amino acids are classified into 7 major classes.

   1. Amino acids with aliphatic side chain. They are also called as *aliphatic amino acids*. They are

glycine, alanine, valine, leucine and isoleucine. Valine, leucine and isoleucine are called as branched chain amino acids.

2. Amino acids with side chain containing hydroxyl groups. They are also called as hydroxy amino acids. They are serine and threonine.
3. Amino acids with side chain containing sulfur atoms. They are also called as sulfur containing amino acids. They are cysteine, methionine and cystine.
4. Amino acids with side chain containing acidic groups or their amides. They are also called as *acidic amino acids*. They are aspartic acid, aspargine, glutamic acid and glutamine.
5. Amino acids with side chain containing basic groups. They are also called as *basic amino acids*. They are arginine, lysine, hydroxy lysine and histidine.
6. Amino acids containing aromatic rings. They are also called as *aromatic amino acids*.They are phenylalanine,tyrosine and tryptophan.
7. Imino acids.They are proline and hydroxy proline.

II. Amino acids are also classified according to the reaction in solution or charge. They are categorized in 3 classes, acidic, basic and neutral amino acids. Acidic amino acids areaspartic acid, glutamic acid. Basic amino acids are arginine, lysine and histidine. Rest of the amino acids are neutral amino acids.

III. Another classification of amino acids is based on the number of amino and carboxyl groups present in the molecule.
Example: Mono-amino mono-carboxylic acid (Glycine), Mono-amino dicarboxylic acid (Glutamate)

IV. Amino acids are also classified according to their nutritional importance. Nutritionally amino acids are classified into.

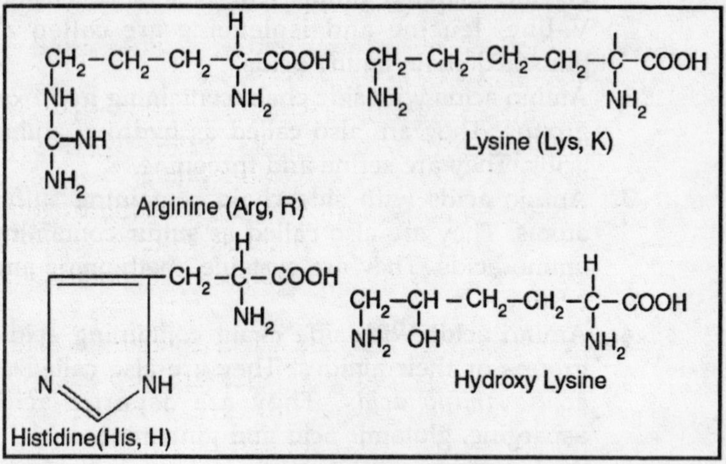

Fig. Basic Amino Acid

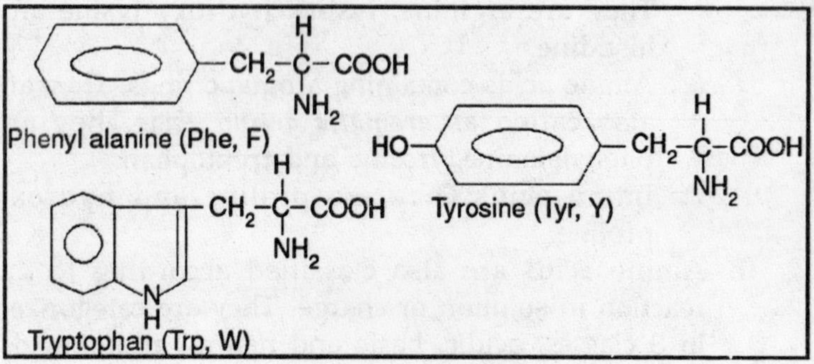

Fig. Aromatic Amino Acid

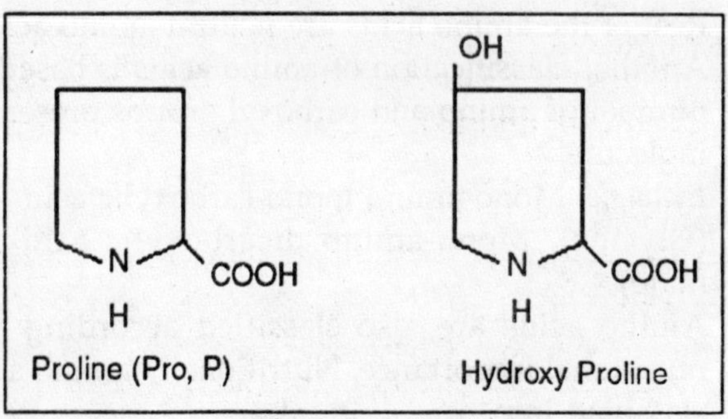

Fig. Imino Acid

(*a*) *Essential amino acids:* These amino acids are not synthesized in the body and hence they have to be obtained from the diet. They are also referred as indispensable amino acids. They are methionine (M), arginine (A), tryptophan (T), threonine (T), valine (V), isoleucine (IL), leucine (L), phenyl alanine (P), histidine (H) and Lysine (L). Together they are remembered as (MATTVILLPHLY). Sometimes histidine and arginine are referred as semi-essential because body synthesizes these amino acids to some extent. Lack of essential amino acids in the diet gives rise to growth failure.

(*b*) *Non-essential amino acids*: These amino acids are synthesized in the body. They are alanine, glycine, serine, tyrosine, glutamate, glutamine, aspartate, aspargine, cysteine and proline. They need not be present in the diet.

Rare Amino Acids or Unusual Amino Acids

These are the amino acids that are not found in proteins but play important roles in metabolism.

Examples

- Ornithine, citrulline and arginino succinic acid of urea cycle.
- α-alanine is part of co-enzyme A.
- Taurine is part of bile acids.
- γ-aminobutyric acid is a neurotransmitter.
- Mono- and di-iodotyrosine are precursors of thyroxine.
- Pantothenic acid is a water-soluble vitamin.
- Homoserine is an intermediate of methionine catabolism.
- *Homocysteine:* It is also an intermediate of methionine catabolism. It is a atherothrombogenic agent. It triggers platelet adhesion. Hence, it is considered as a risk factor for development of coronary artery disease (CAD).

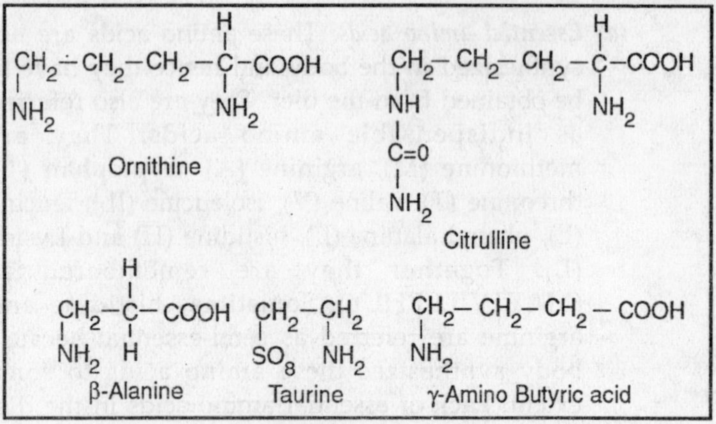

Fig. Unusual Amino Acid

- *S-allylcysteine sulfoxide:* It is an amino acid obtained from garlic. It has many therapeutic effects. It is commonly called as alliin.

PROPERTIES OF AMINO ACIDS

- *Optical isomerism:* All the amino acids except glycine have at least one asymmetric carbon atom because of this they exhibit optical isomerism. Presence of single asymmetric carbon atom gives rise to two optical isomers. One isomer is the mirror image of the other isomer. If a carbon atom is linked to four different groups through covalent bonds then it is called as *asymmetric carbon*. The two mirror images of amino acid serine are L-serine and D-serine. Further, the optical isomers of amino acids are optically active. They are capable of rotating plane polarized light. Some amino acids rotate plane polarized to left and some rotate the plane polarized light to right. All the amino acids present in human proteins are L-isomers. D-isomers are usually absent but they are found in some peptide antibiotics.
- *Acid-base or charge properties of amino acids:* Amino acids act as acids and bases. So they are called as *ampholytes* or *amphoteric substances*. Acids are those

compounds that give protons on dissociation. Bases are those compounds that combine with protons.Bases are also called as alkalies. Proton concentration is quantitatively expressed as pH. It is defined as negative logerithem of proton or H^+ or hydrogen ion concentration.

$$pH = -\log\left[H^+\right] = \frac{1}{\log\left[H^+\right]}$$

The pH scale extends from 1 to 14, which corresponds to hydrogen ion concentration of 1M to 1 × 10^{-14} M. The pH 7.0 represents neutrality pH values less than 7 represents acidity or acids and pH values above 7 refers to bases or alkalinity.

Some common acids are hydrochloric acid (HCl), sulphuric acid (H_2SO_4) and bases are sodium hydroxide (NaOH) and potassium hydroxide (KOH). Further acid is neutralized by base and vice versa.*Function of an amino acid as acid:*

$$R-NH_3^+ \xrightarrow{\textit{Addition of alkali}} R-NH_2 + H^+$$

$$\text{As base:}\quad R-COO^- \xrightarrow[H^+]{\text{Additonal of acid}} R-COOH$$

So, amino acids have two ionizable groups (–COOH, NH_3^+). The –COOH is several times more easily dissociates than –NH_3^+. At neutral pH both groups are ionized, i.e., the carboxyl group exist in dissociated form where as amino group exist as associated form.

This doubly charged molecule of amino acid containing positive and negative charges is called as zwitter ion. It is electrically neutral so it does not move in an electrical field. The other two forms are anionic form and cationic form.

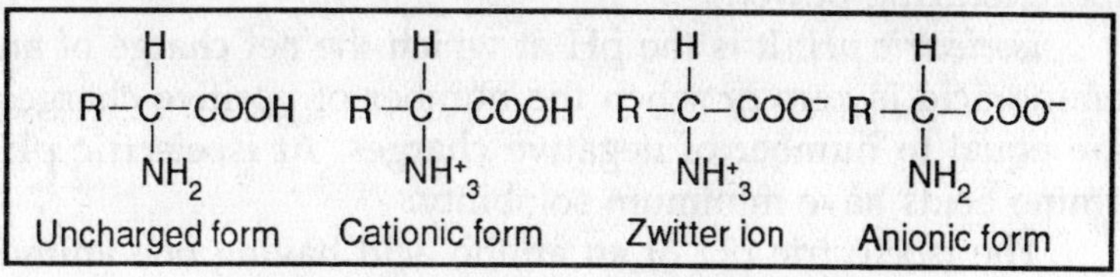

Fig. Different Form of Amino Acids

The charge of an amino acid always depends on the pH

of its surroundings. In other words, the charge of amino acid is altered by changing pH of its surroundings.

This property is exploited for the separation of amino acids. In strong acidic conditions (pH < 2) the –COOH remains undissociated.

When the pH is raised at pH of about 3 the proton from the –COOH is lost –COO– is generated. This is called pK of acid group because at this pH dissociated (–COO–) and undissociated (–COOH) species are found in equal amounts.

Similarly, if the pH is increased to 10, the amino group (–NH_3^+) dissociates to –NH_2 group. This pH is called the pK of amino group of amino acid because at this pH associated (–NH_3^+) and dissociated (–NH_2) species are present in equal amounts.

Therefore, an amino acid has two pK values corresponding to the two ionizable groups. pK values indicates strength of each group. Further an amino acid exist as zwitter ion at neutral pH and as cation at acidic pH and as anion at basic pH.

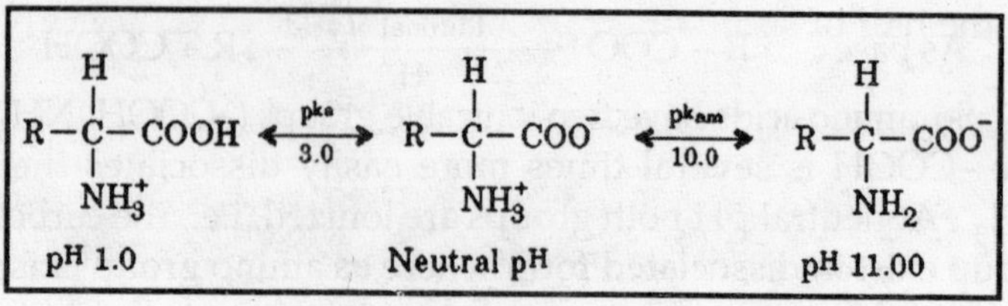

Fig. p^H Influence on Amino Acid Charge

Example: For alanine, pKa is 2.4 and pKam is 9.7 (K is dissociation constant), the low pK value of –COOH indicates more ionizing power.

Isoelectric pH: It is the pH at which the net charge of an amino acid is zero or when the number of positive charges are equal to number of negative charges. At isoelectric pH amino acids have minimum solubility.

The isoelectric pH of an amino acid having one amino group and one carboxyl group is equal to the arithamatic mean of pKa and pKam values.

$$\textit{Isoelectric pH or } p^I = \frac{p^{ka} + p^{kam}}{2}$$

When vaules are substituted, isoelectric p^H of alanineis

$$p^I = \frac{2.4 + 9.7}{2} = 6.50$$

For most amino acids pI is close to 6.0. The situation differs for amino acids having more than two ionizable groups. For example, glutamate is dicarboxylic acid so it can have three pK values (two for carboxyl groups and one for amino group). Similarly, the basic amino acid lysine can have three pK values (two for amino groups and one for carboxyl group). In these cases, a different formula is used to obtain isoelectric pH. For acidic amino acid like glutamate the isoelectric pH is equal to the half of sum of two pK values of acidic groups.

$$p^I = \frac{p^{kal} + p^{ka2}}{2} = \frac{2.2 + 4.3}{2} = 3.25$$

For basic amino acid like lysine the isoelectric pH is equal to the half of sum of two pK values of amino groups.

$$p^I = \frac{p^{kal} + p^{ka2}}{2} = \frac{2.2 + 4.3}{2} = 3.25$$

- *Buffering action of amino acids:* Buffers are salts of weak acids and they resist change in pH when acid or alkali is added. Since amino acids are ampholytes they act as buffers. However, the buffering action of amino acids in the blood is insignificant because of their low concentration.
- *Ultra violet light (UV) absorption of amino acids.* Amino acids do not absorb visible light.

Aromatic amino acids absorb ultraviolet light. Tryptophan absorb ultra violet light at 280 nm. The ultra violet light absorption is also exhibited by proteins containing tryptophan.

Hence, it is used for quantitative estimation of proteins and to study folding of protein molecules. Phenylalanine and tyrosine also absorb light in ultra violet region.

PEPTIDES

- Peptides consist of 2 or more amino acid residues linked by peptide bond.
- A peptide bond is formed when carboxyl group of an amino acid react with α-amino group of another amino acid.. Peptide bond formation between two amino acids is always accompanied by loss of one water molecule. Further, peptide and proteins contain an amino (N–) terminus and carboxy (C–) terminus.
- A peptide or protein is named starting with N-terminal amino acid and usually the N-terminal is located on the left hand side.

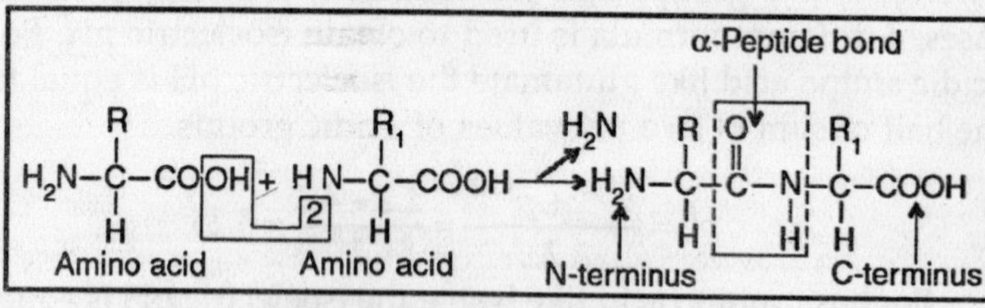

Fig. Peptide bond information

- Animal, plant and bacterial cells contain wide variety of low molecular weight peptides (2-10 amino acids residues) having profound biological functions.

DIPEPTIDES

A dipeptide consist of two amino acid residues and one peptide bond.

Aspartame

Structure: It consist of aspartate and phenylalanine (Aspartyl phenylalanine, Asp-Phe). It is present in African berry.

Function

It is a sweetening agent.

Tripeptides

A tripeptide consist of three amino acid residues and two peptide bonds.

Glutathione

Structure: It consist of glutamate. Cysteine and glycine. In glutathione, γ-carboxyl group of glutamate is involved in peptide linkage with cysteine hence it is named as γ-glutamyl cysteinyl glycine (Glu-Cys-Gly, G-SH).

Functions

- It act as reducing agent in all cells. It assumes dimeric form on oxidation.It is responsible for the maintenance of –SH groups of proteins in reduced form.

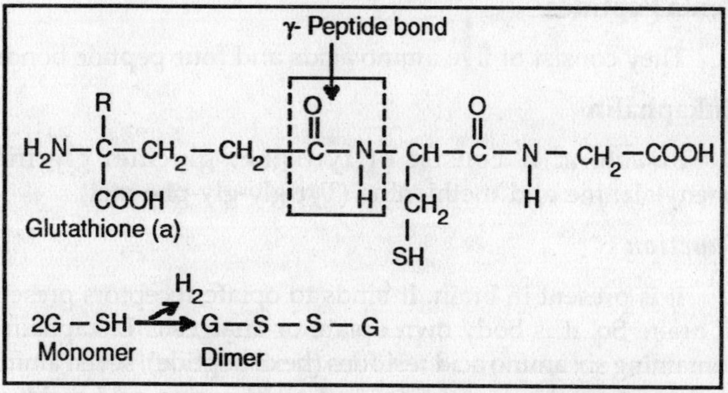

Fig.(a)Structure of glutathione (b)Oxidation of glutathione

- It participates in the removal of H_2O_2 in erythrocytes.
- It is required for removal of toxins from body.
- It is involved in release of hormones.
- It protects body proteins from radiation effects.
- It is involved in cellular resistance to anticancer agents.
- Glutathione regulates telomerase activity and of the cell cycle.
- Glutathione is involved in modulation of apoptosis.

Thyrotropin Releasing Harmone (TRH)

Structure: It consists of glutamate, histidine and proline. It is an unusual tripeptide with blocked N and C terminals.

Function

It is a hormone secreted by hypothalamus.

Chemotactic Peptide

Structure: It consists of N-Formyl methionine, leucine and phenylalanine (f met-leu-phe). Its N-terminal contains formyl (–CHO) group.

Function

It is present in leukocytes. It plays an important role in chemotaxis.

Penta Peptides

They consist of five amino acids and four peptide bonds.

Enkaphalin

Structure: It consist of tyrosine, glycine, glycine, phenylalanine and methionine (Tyr-gly-gly-phe-met).

Function

It is present in brain. It binds to opiate receptors present in brain. So, it is body own opiate or analgesic. Enkaphalins containing six amino acid residues (hexa peptide), seven amino acid residues (hepta peptide) and eight amino acid residues (octa peptide) are also found in brain.

Other noteworthy peptides are:

- *Angiotensin II:* It is an octa peptide, found in lungs and other cells. It is a powerful vaso constrictor and raises blood pressure.
- *Bradykinin:* It consist of nine amino acid residues (Nona peptide). It is a powerful vasozdilator and anti inflammatory.
- *Oxytocin I:* It is also a nona peptide. It stimulates uterus contraction.
- *Vasopressin: A nona peptide produced by pituitary gland:* It has a disulfide bridge. It is also known as antidiuretic harmone (ADH). Angiotensin I and

Kallidin are examples for decapeptides containing ten amino acid residues.

CYCLIC PEPTIDES

- They differ from normal peptides.
- In these peptides N-terminus and C-teminus are linked by peptide bond resulting in cyclization of peptide.
- An antibiotic gramicidin-S is a cyclic peptide. It consist of ten amino acids. So gramicidin-S is a cyclic decapeptide. Further it contains D-Phenyl alanine which is usually absent in life forms.
- Tyrocidin is another cyclic decapeptide.

TOXIC PEPTIDES

- Some peptides act as toxins.
- α-amanitin is a bicyclic octapeptide present in a particular variety of mushrooms. It is extremely toxic to humans.
- It is responsible for mushroom poisoning cases around the world.
- When the mushrooms are consumed it causes pain in the gastrointestinal tract, vomitting, diarrhoea and nausae.
- Death occurs within a week due of impairment of liver and kidney functions.

CYCLOTIDES (CYCLIC PEPTIDES)

In some peptides disulfide bonds are more. These disulfide bonds create knot with in the molecule. Two disulfide bonds and their connecting back bone segment form ring. They are known as cyclotides. These cyclic peptides show diverse actions. Some are anti HIV, anti bacterial and insecticidal agents.

Some examples are given below:

- *Sunflower trypsin inhibitor (SFTI):* It is smallest circular peptide with just 14 amino acids. It is an enzyme inhibitor.

- *RTD-1:* It is a circular peptide present in leucocytes. It is a defensin. It consists of only 18 amino acids.
- *Microsin:* It is 21 residue cyclic peptide. It is produced by E. coli. It is an antibiotic.

Chapter 7

Chemistry of Antibiotics

The treatment of pain and disease is one of the most important goals of humankind. Since ancient times people have been using potions, natural products and even the dust of mummies for the treatment of health problems.

The healing effects of remedies were often ascribed to spirits and mythical entities, but some of the herbal preparations did possess curative properties.

In the 1800's scientists began to investigate potions to determine what chemicals were present that could cause the observed healing.

Thus, the early days of medicinal chemistry began with the study of naturally occurring materials that were effective in treating human disorders. The studies were tedious and required much sample purification and structure determination at a time when instrumental methods of analysis were unavailable. Also, screening methods for chemical efficacy against disease had to be developed so that humans were not used as trials.

ANTIBIOTICS

Modern therapy by chemicals, chemotherapy, is attributed to Paul Ehrlich (Germany, Nobel Prize in 1908) who synthesized the first antibiotic, an arsenic compound patterned after an azo dye that he found to stain a microorganism selectively.

The compound first called compound 606 was introduced commercially in 1910 as a treatment of syphilis under the name of Sylvarsan.

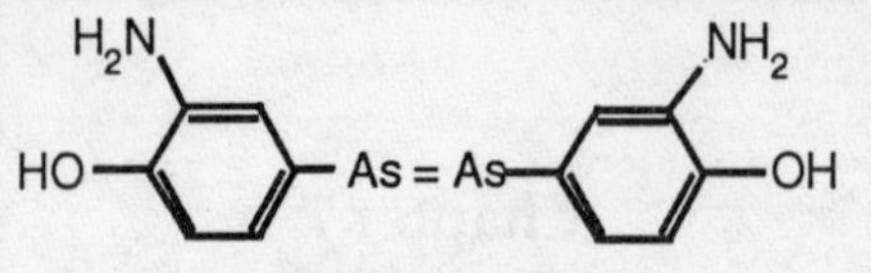

Arsphenamine
(Sylvarsan)

Many compounds were investigated over the next 50 years as antibacterial agents. Some structures are very complex and necessitated the development of new synthetic organic reactions in order to prove their structures.

An intense effort to develop medicinal agents against malaria was required because many soldiers were afflicted with the disease.

Quinine, discovered active against malaria in 1990, was reasonably toxic, and other agents were required. Chloroquine was synthesized and found to be highly effective against malaria.

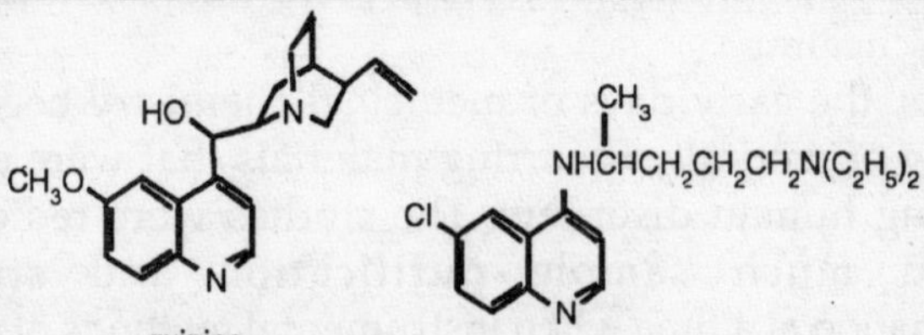

quinine
Isolated from Cinchone Bark

Chloroquine

Careful studies of the antibacterial compound known as Prontisil, a dye that strongly stains proteins, showed that it was metabolized to sulfanilamide, a more effective antibiotic.

Prontosil is therefore termed a prodrug as it is not the substance responsible for the biological effect but it does produce the effective drug.

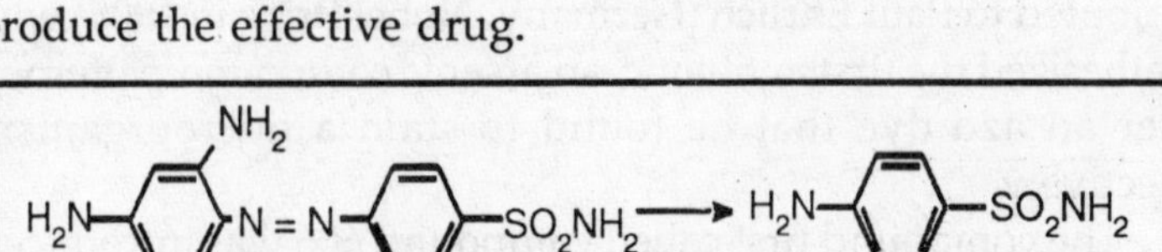

Prontisil

Sulfanilamide

These studies led to the investigation of compounds containing the sulfonamide function as potential antibiotics, and led to a number of very effective antibiotic agents. Many of these compounds are sulfanilamides that have an organic group in place of one of the hydrogens of the sulfonamide. Sulfathiazole proved to be a highly effective antibiotic.

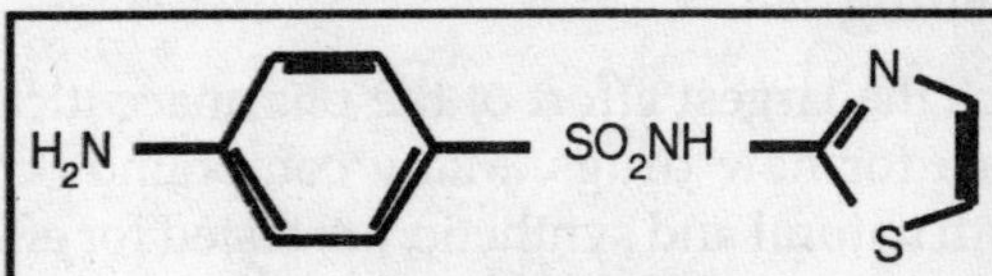

Sulfathiazole

As the search for better antibiotics continued, a revolutionary drug was developed by H. W. Florey and E. Chain in 1938. Penicillin was produced by microorganisms of mold and was found to inhibit the growth of other microorganisms. Penicillin was so important in the treatment of bacterial infections that many research groups from all over the world combined their research efforts in the isolation, purification, testing and synthesis of penicillin drugs. The structurally related cephalosporin antibiotics were also discovered during this time.

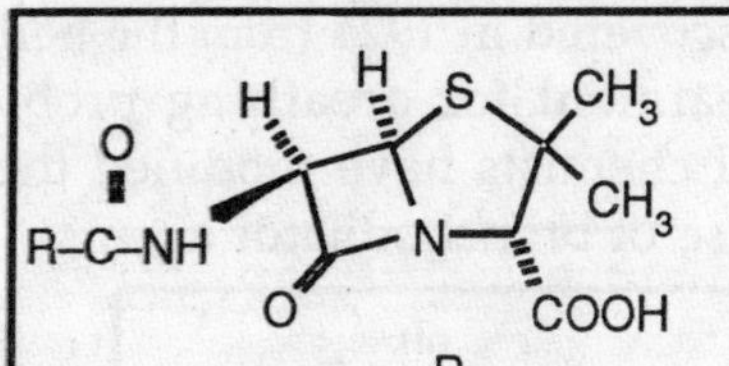

R

Penicillin F $CH_3CH_2CH = CHCH_2$

Penicillin G $PhCH_2$—

Penicillin V $PhCH_2$–

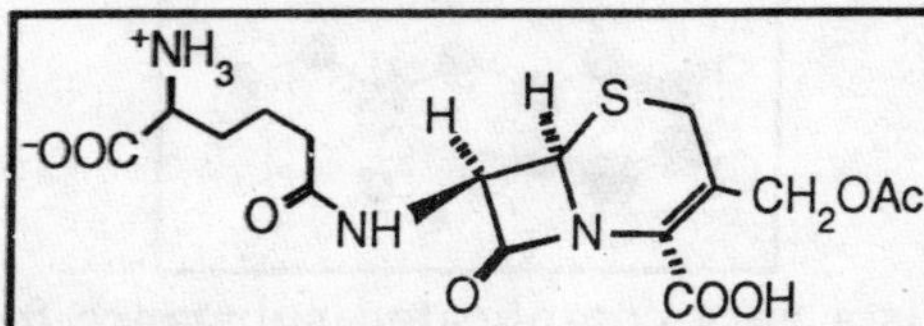

METHODS OF DRUG DISCOVERY

The discovery of new drugs is exemplified in the above sections on antibiotics and analgesics. The procedure is much more refined now than nearly a hundred years ago, but the steps in discovering new drugs or lead compounds to drugs are still similar.

Drug Screening

Perhaps the largest effort of the pharmaceutical industry is the search for new drugs. Many compounds from many sources, both natural and synthetic, are tested for effectiveness against many types of medical problems. Screening requires many involved tests in both test tubes and in live subjects. Analysts need to be very astute in their observations in order not to miss a new drug or to improperly evaluate an effective drug. Once a compound is found to be an effective medicinal agent, then it is the lead compound. New compounds are synthesized in the laboratory in order to find perhaps a better medicine or one with less toxicity. The process takes many years of intense work. Some examples of compounds found in natural sources have been seen as the antibiotics or the analgesics. Scientists then follow up on the lead compounds and attempt to synthesize better compounds. An older drug, ephedrine, was discovered in 1924 from the herbs of Chinese medicine as a treatment for breathing problems such as asthma. Medicinal chemists have modified the structure to provide a wide base of bronchodilator agents.

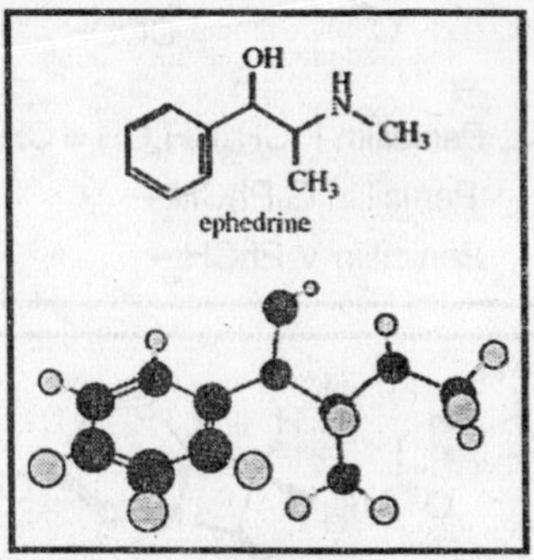

Taxol, is a very complicated substance found in the

California yew tree bark. It is a very important new anti-cancer treatment, but it is in very short supply. Chemists are very active in the synthesis of taxol and related compounds in order to produce more effective and more available materials.

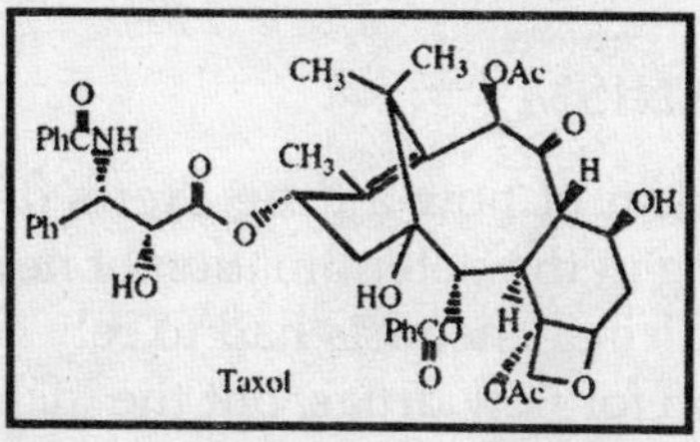

Taxol

Other interesting drugs have been synthesized in the laboratory but their effectiveness as medicines was not discovered immediately. AZT, an important agent against HIV, was first prepared many years ago but was later screened as an anti-HIV agent because the AIDS virus was discovered after AZT was prepared.

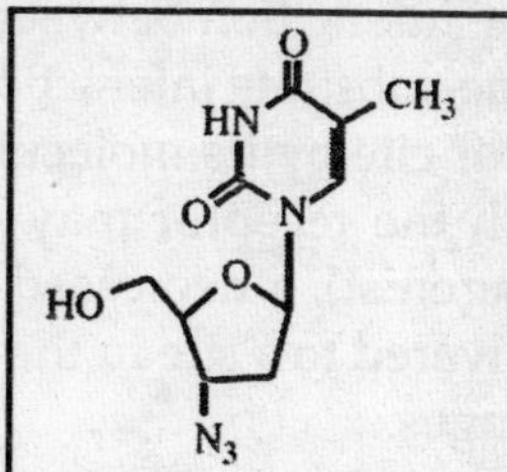

3"-adido-3'-deoxythymidine

AZT

Librium, an important tranquilizer and somewhat abused drug, was tested two years after its first laboratory synthesis because it was not originally thought to be an important structure. As a lead compound Librium has led to several other important tranquilizers.

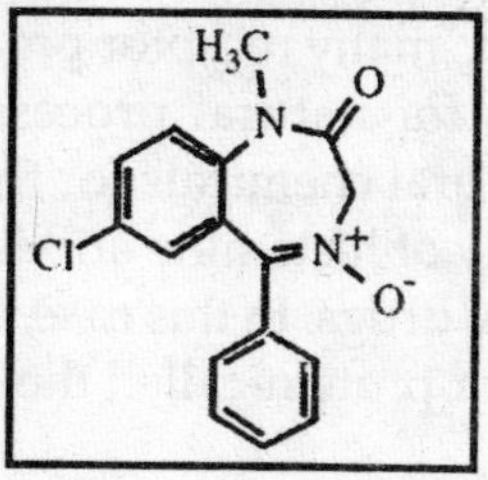

The examples shown above serve to illustrate the beginning of a process. Many other cases are known where drug discovery is based on unusual natural products, or materials based on herbal remedies, or from fortuitous accidents.

DRUG MECHANISM

The mechanism of how a drug exerts its effect is a very significant finding in the determination of new and improved drugs. Early medicinal chemists had to rely on screening and drug modification for new drugs, but the inclusion of modern mechanistic theories into drug searches advances the field considerably. However, the fortuitous discovery of a drug without much knowledge of how or why it works still plays an important role in medicinal chemistry.

A major finding in drug mechanism is that many drugs react with a receptor protein that occurs in cell walls or in the cytoplasm of cells. The interaction between drug and receptor is similar to the enzyme-substrate interaction in which the drug must fit into a receptor cite by its molecular shape. The drug in its interaction with the receptor may cause an enhanced biological response (agonist), a decreased biological response (antagonist) or be delivered to a site in the cell where the drug effects beneficial changes.

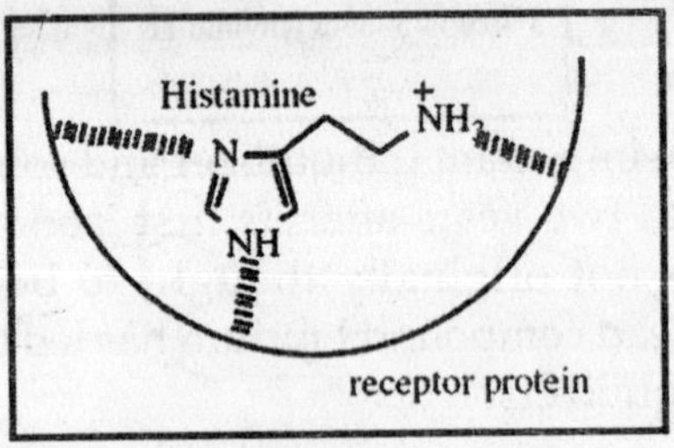

The body contains many receptor proteins that are present as receptor proteins for natural processes. The drug must compete with the natural chemicals for the protein in order to be effective. The study of histamine and its receptor permitted rational design of new drugs. In this case, histamine was found to bind with a receptor protein-called the histamine-2 receptor

- as shown below. New drugs were deigned that would mimic histamine binding but would not permit histamine to bind. The effects of these types of drugs were to act as antihistamines.

The interaction of the antihistamine promine with the histamine receptor is shown below.

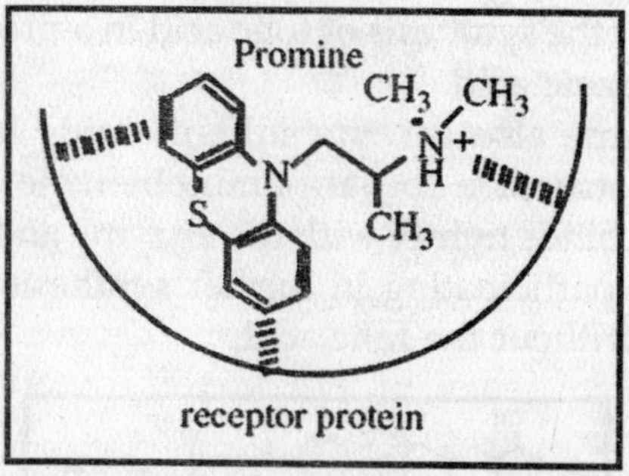

A novel approach to the treatment of breast cancer is the use of agents that bind with the estrogen receptor protein. Mammary cancers are known to have a buildup of estrogen receptor protein and estradiol, the primary female hormone. An estradiol analog with a radioactive fluorine atom in the 16a position competes very strongly for the estrogen receptor protein.

Thus the mammary cancer becomes labeled for diagnostic purposes and the radiation produced from the 18F can destroy the nearby cancer cells. The radioactive fluorine compound is called a radiopharmaceutical.

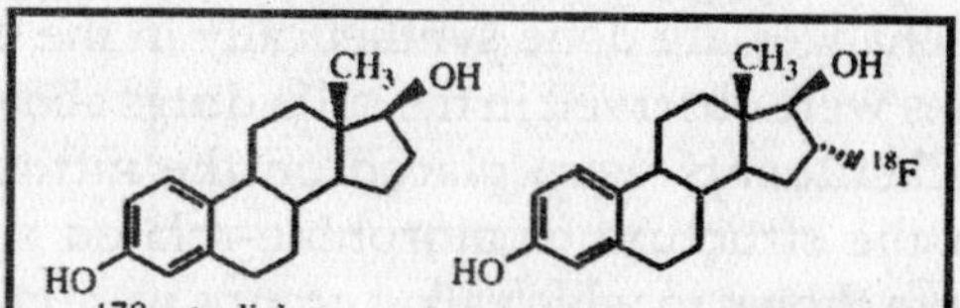

17β-estradiol 16α-[18]Fluoro-17β-estradiol

A very important type of drug is one that interacts with an enzyme. The enzyme is fooled into using the drug as a natural substrate, but the drug does not continue the function of the enzyme and its activity is inhibited. Such drugs are called mechanism-based suicide inhibitors.

The reaction mechanism is used to design the drug and the enzyme commits suicide when it accepts the drug. An example of this behaviour is found with many antibiotic drugs that inhibit the growth of bacterial cells. Bacterial cells must synthesize folic acid whereas humans obtain folic acid from food. An enzyme named dihydropteroate synthetase is responsible for the synthesis of folic acid in a process that uses paraaminobenzoic acid.

The enzyme also accepts sulfanilamide because of its similarity in structure to para-aminobenzoic acid, but the sulfanilamide binds tightly with the enzyme and prevents the enzyme from participating in further synthesis of folic acid. The cell dies without the folic acid.

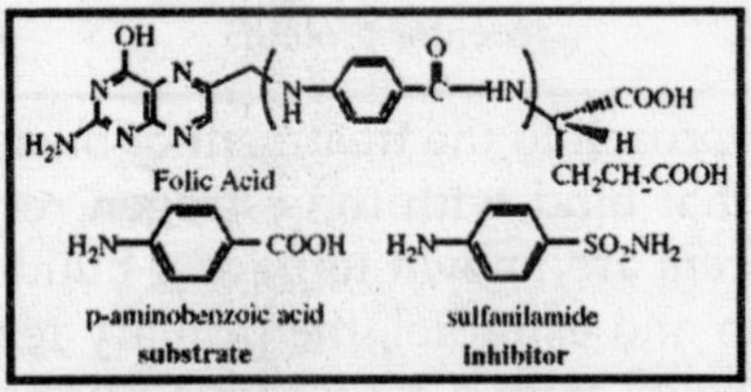

The mechanism of action of most drugs involves receptor binding and enzyme inhibition. The study of enzymes and the structure of active sites are thus major efforts in medicinal studies.

DRUG ALTERATION

Once a lead compound with medicinal properties is discovered changes are done synthetically in the compound. Such changes were observed in the sulfa drugs above in which different substituents were placed on the nitrogen. Many changes in the structure of morphine-related compounds (opiates) were shown in which alkyl groups were changed and rings were removed systematically.

The important finding is the lead compound, but very often the altered drug shows better properties or less toxicity. Another alteration of drugs is the incorporation of fluorine. Fluorine atoms allow the original structure to retain its molecular size because fluorine is a small atom.

Also, the fluorine is very electronegative and thus introduces polarity into the molecule. The change can sometimes be very beneficial in drug discovery. Steroid compounds are often used as anti-inflammatory agents and many with incorporation of fluorine atoms are effective at lower doses, as is the case with fluorocortolone shown below.

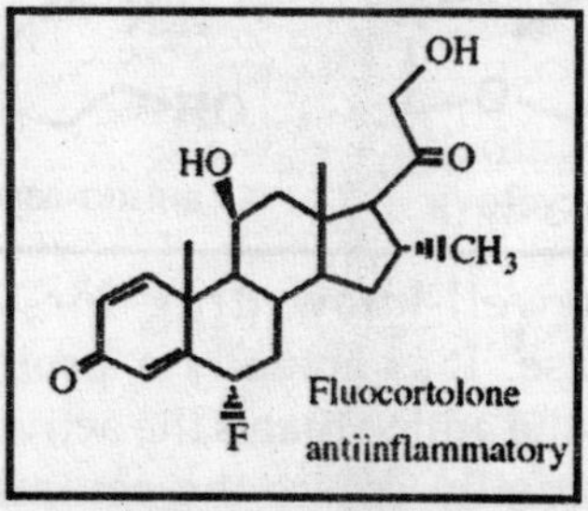

Synthetic incorporation of fluorine into other drug systems is also well known and very effective. Fluorazepam is an often prescribed relative of valium.

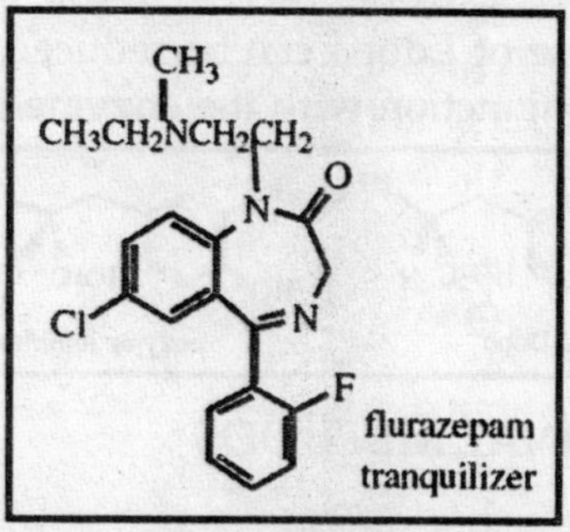

Once a drug is discovered many experiments are conducted to determine the pharmokinetics of the drug. These studies show how the drug is delivered to the site of action, how the drug is distributed throughout the body and how the drug is metabolized.

The results of the studies sometimes provide clues as to how the drug could be altered to improve its pharmokinetic properties and effectiveness.

Biodistribution studies of the drug acyclovir, used in the treatment of herpes simplex virus, show that it is only absorbed about 15% when taken orally. A better absorption is found when the carbonyl group is replaced by an amino group.

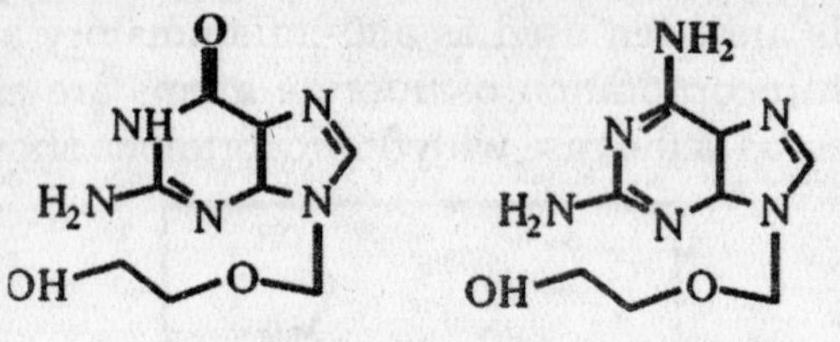

Acyclovir amino-acyclovir

L-Dopa is a well-know drug for the treatment of Parkinson's disease. It is actually a prodrug because it is decarboxylated to the amine that is the active drug. Only about 1 % of L-dopa actually enters the central nervous system because it is decarboxylated by enzymes in the liver.

A solution to improving the uptake of L-dopa is to add another drug that inhibits the decarboxylation enzyme thus permitting more L-dopa to reach the active site. The administered dose of Ldopa can be reduced more than 50 % when used in conjunction with the enzyme inhibitor.

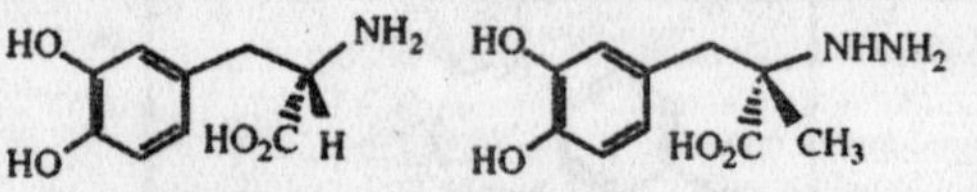

L-Dopa enzyme inhibitor

COMPUTATIONAL METHODS

Hansch Equation

A computational method to predict drug effectiveness based on physical properties instead of molecular structure was introduced by Hansch in 1960. In this method parameters on the hydrophobic (water dislike) character of the compound (P), the hydrophobic contributions of the substituents (p), electronic characteristics of the substituents (s, known as Hammett constants) and steric factors (Es, known as Taft constants) are correlated with drug effectiveness. The general Hansch equation is shown below.

$$\text{drug effectiveness} = k_1 \log P + k_2\, p + k_3\, s + k_4\, E_s + k_5$$

The k values indicate the significance of each term. When applied to a number phenethylamine compounds related to ephedrine, the drug effectiveness was found to depend only on substituent (X and Y) hydrophobic and electronic properties.

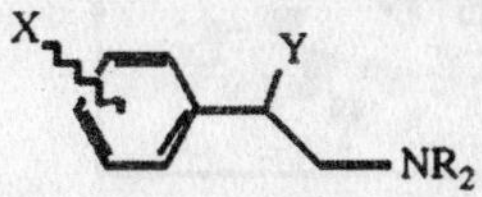

$$\text{drug effectiveness} = 1.2\,\pi - 1.6\,\sigma + 7.9$$

The equation says that compounds with highly hydrophobic (positive p values) and electron donation (negative s values) will prove to have greater effectiveness.

The Hansch equation for the above compounds is relatively simple, but for some drugs a very complicated relationship is observed. The Hansch study of drugs requires that many compounds be evaluated and correlated with the various parameters, not a simple task.

Molecular Modeling

Modern high-speed computers have the capability of carrying out many complicated calculations in a very short time. The structural parameters for a specific protein receptor (or enzyme) can be placed in a computer along with structural parameters for a drug molecule (real or hypothetical) and the computer will calculate and graph the fit of the drug with the protein active site.

The molecules may be manipulated in three dimensions to enhance the view and show the best fit of the "docking" of the molecules. Such molecular modeling can be used to predict the structure and effectiveness of new medicinal agents. Below is a model of captopril, a blood pressure medication.

This model is produced by drawing the structure with a ChemDraw programme, and placing it into a Chem 3D programme.

The structure is presented in colour and may be

manipulated in 3D with the programme. Examination of the structure when one has the shape of the enzyme active site can give information about the desired location of substituents.

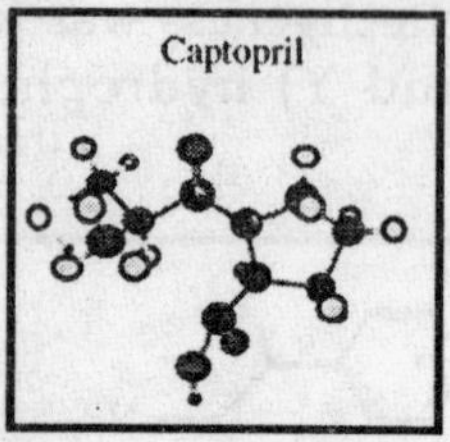

Successful Drugs

Medicines have been developed to treat almost every disease. The development of medicinal agents takes many years of intense work by many scientists skilled in different fields. A few of the commonly prescribed medications are shown below.

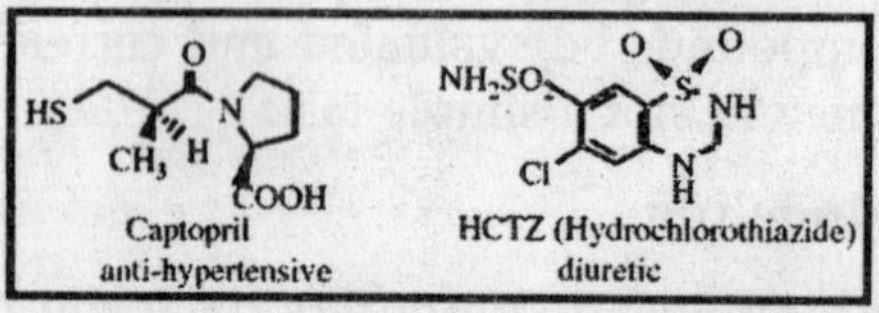

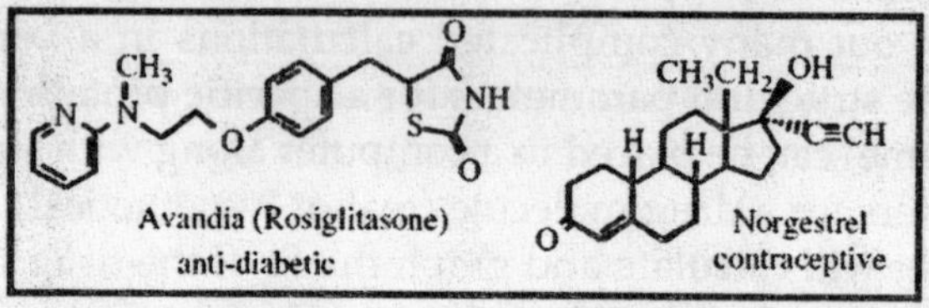

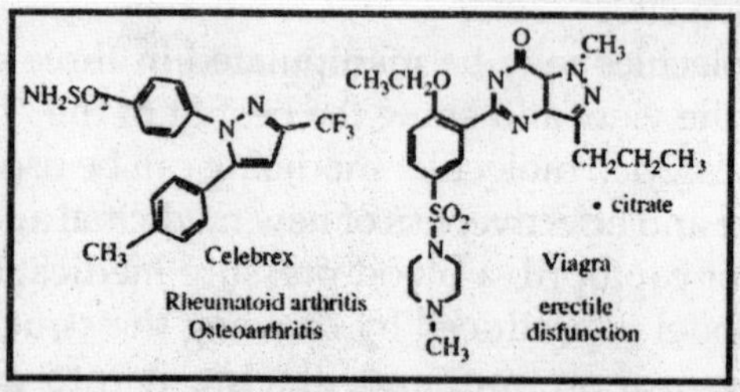

Although various treatments for diseases have been used for centuries, the early history of chemotherapy, chemical treatment of disease, started with *Paul Ehrlich in the 1900' and his discovery of antibiotic compounds.*

Many antibiotics with a wide variety of applications and structures are now known. Drugs known as *prodrugs* that release the true drug have become common through the studies of the mechanism of drug action.

The *opiate family* of drugs is used in the treatment of pain. Although, many are highly addictive, the systematic study of their mode of action has led to a modern approach for drug discovery. Now known drugs are modified according to structural features thought to be responsible for the drug activity.

The wellknown *β-phenethyl amine structural unit of opiates* is identified in many structures with known chemotherapeutic properties. Drugs are often *discovered by extensive screening tests* of materials synthesized in the laboratory or obtained from natural sources.

Studies are then conducted to gain information on the *mechanism of the drug action.* Drug *interactions with enzymes* as substrates or inhibitors then lead to the structural features required in designing new drugs.

The drug structure is modified synthetically according to the drug mechanism in order to provide new drug molecules. *Computational methods* are used to show the correlation of drug structure with physical and biochemical properties of the drug, and to determine the efficacy of the drug.

The computations lead to drug modifications for changing the required properties. Computers are often used to *model drug chemistry.* Computer programs will allow the observation in *three-dimensions of a drug interaction with a protein, or permit the determination of the stereochemistry of the drug.*

Index

A

Acetals 107, 108, 119
Achromobacter 138
Acidic 110, 190, 193, 196, 213
Aconitase 144, 145
Active Site 284, 286, 287, 289
Adaptogenic Plants 249
Adhesion 129, 130
Aglycone 230, 238, 245, 248
Albumin 4, 5, 14, 15, 16, 17, 18, 25, 36, 37, 38, 43, 44, 45, 47, 81, 93, 189
Alcoholic Glycosides 230
Alpha Helix 26
Amino Acids 19, 267
Amphipathic 27
Anaerobic Respiration 150, 162
Angiotensin 31, 32, 33, 263, 274
Animal Foraging 246
Anomeric Forms 106
Anthraquinone Glycosides 230
Antibiotic Inhibition 135
Antiporters 56
Apoptosis 47, 48
Arachidonic Acid 44
Aromatic 110, 163, 175, 205, 213
Ascorbic Acid 171, 172, 174, 192, 197, 198, 199, 200, 209
Asparagine 21, 127, 128, 134
Aspartame 272
Atherosclerosis 34, 41, 43

B

Bacillus Subtilis 158
Bacterial Plasma 57
Bacteroides 138
Beta Barrel 28
Beta Sheet 28
Beta Strand 28
Bilayer Membranes 181
Bioactivities 247
Biochemical Mechanism 243
Biosynthesis 131, 146, 161, 245
Biotin 171, 174, 177, 189, 190, 191
Brucella 138

C

Calcineurin 45, 46, 48, 49
Calcium Channels 52
Capnocytophaga 138
Carbohydrate Metabolism 141
Carbohydrates 98, 99, 115, 121, 123, 124, 125, 126, 127, 128, 138, 139, 140, 141, 142, 148, 168, 170, 184
Cardiac Glycosides 232, 237, 238, 242
Cardiovascular 32, 83, 84, 85
Catalyzed Transport 55

Cellobiose 111, 112
Cellular Distributions 37
Cellular Respiration 166
Cellulose 99, 111, 112, 113, 114, 115, 121, 125, 126, 171, 222, 223, 224
Channel Structures 53
Chaperone-Mediated 75
Chaperones 70, 72, 73, 74, 75, 76, 78, 79, 80, 81, 83, 86, 96
Chemical Nature 172, 230, 233
Chemokines 89
Chemotactic Peptide 274
Chemotherapy 277, 288
Choline 18, 142, 171, 172, 174, 175, 176, 177, 184, 201, 202,
Citrate Synthase 144, 145, 148, 149, 168
Clinical Assessment 176, 208
Computational Methods 286
Conjugated Proteins 86
Coumarin Glycosides 230
Coupled Receptors 50
Cyanogenic Glycosides 230
Cyclic Forms 107
Cyclic Peptides 275
Cystic Fibrosis 69
Cytochromes 134, 150, 152, 153
Cytokines 18, 86, 87, 88, 89

D

Defence 131, 212
Deficiency Syndrome 173, 181, 184, 192, 196, 198, 203
Deficiency Syndromes 173, 184
Dehydrogenase 2, 13, 46, 141, 142, 143, 145, 146, 147, 148, 149, 150, 151, 152, 154, 161, 163, 167, 178, 204, 236, 237
Diphosphokinase 143
Disaccharides 110, 111, 112, 123
Disease 277, 278, 286, 288, 289
Drug Discovery 280
Drug Mechanism 282
Drug Screening 280

E

Electrochemical 51, 55, 56, 57, 58, 59, 156
Electron Transport 141, 149, 154, 160, 165, 187, 212
Electrostatic Bonds 12
Elongation 124
Energy Transformation 142
Enkaphalin 274
Enzyme Inhibitor 286

F

Flagellar Motion 161
Flavobacterium 138
Flavonoid Glycosides 231
Flavoproteins 150, 151, 152
Fluorine 283, 284, 285
Folic Acid 171, 174, 192, 193, 194, 195, 196, 197, 204
Fumarase 147
Fumarate Hydratase 147

G

Gardnerella 138
Gentiobiose 111, 112
Globulins 1, 5, 15, 16, 17, 18, 19, 93, 128, 131
Glutathione 273
Glycone 228, 229, 230, 231, 232, 233, 236, 238, 239, 240, 241, 243, 245, 248, 250, 251, 258

Glycoprotein 17, 86, 127, 129, 130, 131, 132, 133, 134
Glycoside 109, 110, 111, 112, 113, 114, 222, 228, 229, 230, 231, 235, 236, 244, 247, 249, 251, 254, 256, 258, 261
Glyoxylate Cycle 168

H

Hansch Equation 286
Harmone 273, 274
Health Controversy 261
Heart Failure 33, 232, 238
Helix 7, 8, 10, 25, 26, 27, 29
Hemiacetals 106, 111, 119
Hexoses 99, 100, 105, 108, 116, 122
Hormones 130, 133, 170, 181
Human Disease 70
Hydrochloride 172, 176, 180, 182, 203
Hydrogen 279
Hydrogen Bond 113, 140
Hydrophobic Molecules 44

I

Identical 102, 176
Immunoglobulin Family 50
Immunological 131
Immunosuppression 45, 48
Inactivation 42, 43, 175, 179, 182, 185, 188, 191, 193, 196, 200, 202, 212
Inhibitors 31, 32, 33, 43, 48, 93, 131, 263, 264
Inositol 171, 172, 174, 176, 177, 200, 201, 202
Insulin 6, 262
Interferons 88
Interleukins 86
Intracellular Shuttle 86
Ion Channels 51, 53, 54, 65, 96
Ion Selectivity 51
Ionophores 156, 157
Isocitrate Dehydrogenase 143, 145, 161, 167
Isomers 100, 105, 106, 117, 118, 122, 152, 201, 202

K

Ketoglutarate 143, 145, 146, 167
Ketoses 102, 105, 122

L

Lactose 102, 109, 112, 113, 120, 122, 123, 124, 127, 139, 140, 141, 161
Leucine Zipper 30
Lipoic Acid 204
Lipopolysaccharides 137, 138

M

Malate Dehydrogenase 148
Maltose 111, 112, 113, 121, 123, 126, 140
Medicinal Chemistry 277, 282
Membrane Proteins 50
Metabolism 42, 141, 162, 181, 184, 187, 189, 192, 195, 202
Metallochaperones 86
Metalloenzymes 86, 134
Miscellaneous Proteins 70
Modification 282, 289
Molecular Chaperones 70, 71, 72, 73, 75, 76, 78, 79, 81, 82, 83, 96
Moraxella 138
Myosin 91, 92

N

Niacin 171, 176, 177, 186, 187, 188, 189
Nicotinic Acid 171, 174, 178, 186
Nitrate reductase 163, 164
Nitric Oxide 44
Nitrobacter 162, 163
Nitrogen 284
Nomenclature 46, 102, 111, 239, 245, 251

O

Osazone Formation 103
Oxidation 100, 101, 102, 103, 104, 105, 106, 109, 120, 141, 143, 145, 147, 148, 149, 162, 163, 165, 166, 167, 173, 175, 178, 184, 187, 191, 197, 199, 204, 205, 207, 209, 210, 211, 212

P

Pantothenic Acid 183
Pectinatus 138
Penta Peptides 274
Pentoses 104, 116, 121, 122
Peptides 2, 7, 9, 23, 30, 57, 64, 66, 70, 80, 81, 82, 95, 263, 264, 272, 274, 275
Peptidoglycan 135, 136
Pharmaceutical Targets 43
Pharmacokinetics 242
Pharmacological Intervention 82
Phenolic Glycosides 231
Phosphorylation 36, 39, 44, 45, 48, 49, 61, 62, 142, 143, 146, 147, 149, 154, 165, 166, 167
Plasma Proteins 16
Plasma-Membrane 59, 66
Polypeptides 2, 7, 9, 23, 30, 57, 64, 70, 80, 81, 82
Polysaccharides 58, 98, 112, 113, 114, 116, 120, 121, 122, 123, 124, 125, 126, 137, 138, 223, 224, 225, 228
Porin 54
Positive Functions 172, 181, 183, 192, 195, 197, 203
Potassium Channels 51
Protein Chaperones 73
Protein Folding 75
Protein Structure 13
Pseudomonas 138
Pyridoxine 171, 173, 176, 177, 178, 180, 181, 182, 183

R

Rainbow Trout 139, 198
Random Coil 10
Receptor Binding 284
Reproduction 130, 135
Riboflavin 122, 171, 173, 176, 177, 178, 179, 180

S

Sarcoplasmic Reticulum. 59
Selenomonas 138
Sodium Channels 52
Sorangium 138
Steroid Hormone 133
Steroidal Glycosides 232
Steroids 44, 45, 234, 238, 242, 248
Stevia 232, 249, 250, 251, 253, 254, 255, 256, 257, 258, 259, 260, 261, 262
Steviol Glycoside 251, 254, 256

Stoichiometry 100, 148, 155
Structural 99, 103, 106, 107, 108, 113, 116, 125, 126, 127, 129, 135, 188, 201, 212
Structural Analysis 217
Structural Disorder 73
Structural Proteins 91
Succinate Dehydrogenase 147
Succinyl 146, 147, 195
Sucrose 111, 112, 120, 121, 123, 124, 126, 139, 140
Sugar Moiety 241
Sugars 16, 55, 58, 66, 98, 99, 101, 102, 103, 105, 110, 111, 112, 113, 115, 116, 117, 119, 120, 121, 122, 123, 125, 132, 133, 135, 136, 137, 141, 214, 215, 216, 217, 218, 220, 221, 223, 224, 225, 226, 241, 242, 248
Sulfur 145, 151, 162, 164, 167
Super Helix 10
Symporters 51, 56
Synaptic Channels 54
Syndrome 173, 178, 181, 184, 185, 187, 188, 190, 192, 196, 198, 201, 203, 205, 207, 210, 212, 262

T

Termination 123, 208
Tertiary Structure 11, 12, 16, 55
Therapeutic 289
Therapeutic Claims 247
Thiamine 171, 172, 173, 174, 175, 176, 177, 187, 188, 204
Thiamine Pyrophosphate 172, 204
Thiobacillus 138
Thioglycosides 232
Thyrotropin 273
Tilapia 139
Tissue 1, 17, 18, 37, 38, 40, 84, 129, 141, 146, 173, 175, 178, 179, 198, 199, 200, 201, 202, 204, 205, 207, 209, 214, 250
Toxicity 14, 43, 204, 238, 246, 247, 256, 257, 262
Transhydrogenase 161
Transmembrane 36, 37, 50, 53, 54, 56, 57, 58, 60, 62, 64, 66, 69, 166
Transport Proteins 51, 55, 56, 58, 66
Treatment 277, 279, 280, 281, 283, 285, 286, 288, 289
Trehalose 111, 112
Tripeptides 272
Triple Helix 10
Triterpenoid Saponins 248, 249
Tumor Necrosis 47, 88

U

Ubiquinone 150, 151, 152, 163, 164
Ubiquitin 92, 94, 95, 96, 97

V

Veillonella 138
Vibrio 138
Vital 99, 170
Vitamins 2, 66, 130, 169, 170, 171, 172, 179, 187, 191, 203, 204, 205, 208, 210, 211, 212, 263

Z

Zinc Finger 28, 29, 30